R.O.A.M.

Published by Dr Jonathan Royle
Publishing partner: www.magicalguru.com
(c) Dr Jonathan Royle 2015 – This Edition
© Dr Jonathan Royle 2008 original Publication

Book design, layout and production management by www.magicalguru.com
Printed and bound in UK and USA by Lightning Source

FOREWORD – NUMBER ONE

R.O.A.M – Ramblings Of A Madman – Space,

Time, Travel, Evolution, Pyramids, Einstein,

Darwin, Aliens, UFO's, Ghost's, The

Paranormal, Supernatural and Reality Of All

Matter Revealed & Explained

Dr. Wilf S G Archer, PhD; Chartered MCIPD; CMIOSH;

GHR(Reg); NHR.

I have been an avid student of such Laws,

studying Books, Manuscripts, Writings and

Transcripts, ever searching for the

essential truth whether devised by man or

revealed by God.　　　　A truth that embraces all

systems of philosophy, sociology, science

and religion – from Zarathustra to Spencer,

from Jesus to Quisquam, from Old Thought to

New Thought.　　　　Collecting knowledge and

understanding devoted to the study of man

and the Art of Life.　　　　Guided by intuition

and reason, examining their teachings as

founded on experience and observation and

insisting that the efficacy and value of

every claim be measured by practise and

demonstration.　　　　Over the years I have seen

some wondrous sites and witnessed feats of Magick that would raze the logic towers of academia to rubble.

All the lost cities and sunken treasures matter not more than a beggar's dole when compared with the advancements of mankind. It cannot be denied that the past century has been the most exciting epochs in the history of mankind, and the scientific discoveries of the past 15 years have brought about a rejuvenescence of human thought.   Nor can we belittle the conquests which the scientists claim for the latter half of the last century, among which may be especially named:

a.   The proved uniformity in the Laws of Life.

b.   The true character of disease and DNA manipulation.

c.   The immense possibilities of quantum theory.

d.   The development of renewable energy.

e.   The great advancements in Biological Physics.

f.   The general unity between the Macro and
      Micro forces of the entire Universe.

g.   The possibilities through harnessing the
      speed of light.

All this advancement culminating in what may
be called the Laws of Substance.   But now we
have moved on.   Now we have grown in our
understanding, maturing and moving towards
the very outer limits of our material
Universe.   A new world awaits beyond that
which can be seen.   A world beyond our
physical senses.   A world that exists within
the spaces between the atomic structures.   A
world of wonder, of magic and of confusion.
A world where the mere act of observation
transforms its very existence, a world
moulded by the reality of perception.
However, tread carefully dear reader as
these pages are as yet only a preliminary
map and like the proclamations of the great
adventurers of this world contain many
uncharted pathways.   Remember Columbus who
made the Earth round or Galileo who altered
the orbit of the planets –they changed
nothing but the perceptions of man but that

change of perception changed everything and

the world was never the same again.   First

the pioneer reveals a truth which goes

against traditional thinking which is often

rejected as the ramblings of a madman.   Then

when we can no longer hide from the enormity

of the revelation we realize that the

madness we projected was only a reflection

of our inertia and thus begins our

transformation as we learn to use that truth

to our advancement.   Then the world changes.

As it was so it will always be   - it is us

that change.

Ramblings of a Madman is such a revelation,

a poetic work with such depth that its

profoundity for change is enormous.

We have reached that stage in progress when
our creeds and laws must change,
When our honest thoughts and efforts must
have wider scope and range,
As we stand at this new beginning - freed
from ignorance and strife.
Poised in anxious readiness for the tools to
Change our Lives.

Dr. Wilf Archer, founder of The Mindskills Approach, is the UKs only Double Chartered*, Professional Health Coach. His expertise is in empowering, enabling and equipping his clients with the skills to overcome the problems of chronic ill health, depression, stress, addiction and unreliable performance. His unique style is an advanced form of Educational Intervention and is the only integrated coaching style that focuses on supporting The Individual at a Core Level.

Visit Dr Archer's website for further details http://www.mindskills.co.uk

FOREWORD – NUMBER TWO

R.O.A.M – Ramblings Of A Madman – Space,
Time, Travel, Evolution, Pyramids, Einstein,
Darwin, Aliens, UFO's, Ghost's, The
Paranormal, Supernatural and Reality Of All
Matter Revealed & Explained

My name is Dr. Jonathan Royle and I was born
Plain Alex William Smith on the 13th August
1975 into a showbiz family whilst travelling
with Gandey's Circus.

These days I am most widely known for my
work as a NLP Hypnotherapist & Psychological
Change Consultant and indeed it is in this
capacity that I came to meet the true author
of this amazing book namely T. H.uGoth who I
also call Tom for short!

I call him Tom when we communicate as that
stands for Truth Of Matter and that's
exactly what the contents of this revealing
publication are all about.

So open your mind and trust me if you enjoy the contents of this book half as much as I feel sure you will then please check out my other titles which are available from www.amazon.com and all other good book sellers worldwide including:

*Confessions Of A Hypnotist (Everything You Ever Wanted To Know About Hypnosis But Were Afraid To Ask)

*Confessions Of A Celebrity Psychic (How I Became Rich & Famous As A Fraudulent Psychic Entertainer & Consultant)

*Confessions Of A Showbiz Celebrity (How I Became Rich & Famous As A Showbiz Celebrity)

All of the above having been written and released in my stage name of Dr. Jonathan Royle.

Also you may like to take a look at the book to which I recently advised and wrote the foreword entitled: *The Easy Buddha (by KYIMO)

You Will also most likely find much of interest and value in my huge range of Hypnosis, NLP, Psychology, Mentalism & Related Training DVDS, Audios and courses, more details of which along with details of my success and experience in these industries to date can be found at the following Internet Links:

01) www.hypnotherapycourse.net

02) www.hypnotorious.com

03) www.hypnotictraining.com

04) www.mindpowerprofits.com

05) www.millionaireplan.net

06) www.magicalguru.com

http://www.selfgrowth.com/engine/experts/jonathan_royle.html

ALSO YOU MAY WISH TO JOIN MY FRIENDS LIST'S AT:

MYSPACE = http://www.myspace.com/magicalguru

FACEBOOK = http://www.facebook.com/people/Alex_William_Smith/673582805

BEBO = http://www.bebo.com/AlexS0921

ECADEMY = http://www.ecademy.com/account.php?userid=alexandersmith

STARNOW = http://www.starnow.com/alexsmith13

THE OFFICIAL JONATHAN ROYLE FANCLUB IS AT:

http://www.facebook.com/group.php?gid=32188053824

And To See Dr. Jonathan (Alex Smith) Royle in action both on Stage, Television and at live training seminars and events you should visit:

http://www.youtube.com/user/HypnotherapyCourse

http://www.youtube.com/user/hypnotorious

http://www.youtube.com/user/promoclips

http://www.youtube.com/user/dreameruk

http://www.youtube.com/watch?v=7zNJi1V0zAc

Sincere Thanks & Enjoy The Book

Dr. Jonathan (Alex Smith) Royle

PS: You can also listen online Free to the
original Radio Interview on Paranormal
Palace Radio About the contents of this book
at this link:
www.blogtalkradio.com/paranormalpalace or
www.paranormalpalaceradio.com

The

DIGEST

Of

R.O.A.M.

By

T. HuGotH.

Ramblings of a Madman.

Relating to

Reality of all Matter

Manifested through the chosen prism,

Alex W. Smith.

Dedicated to Reality and Others.

The Dream is the truth of the Dreamer.

To Dream is to Exist.

I

Writing is the stage on which I act, for all
the world to see. I am an actor with words
and I play many characters. The page is
where I perform and when I am not on page I
am my private self. But words have now
fallen. It is a script that I must act. A
great page has been created on which I must
try and give my best performance.

Applause may not be worthy,
And my little act forgot,
But I have braved to stand on stage,
The audience has not.

T. HuGotH.

We are the Dream of the Universe.

II

Reality is a state of mind,

Mind is the reality of thought,
Thought is the reality of itself,
Itself is reality.

AWAKENING.

The enlit first reached as I stood by a tree
at the corner of the large schoolyard. I was
gazing at the imposing red brick 1908
building, with its vast ski slope of tiles
covering the roof of the great hall. It
seemed vast, for I was so little on my first
day. Then I did not know that size is
meaningless. All is relative to the
conscious view point. You can cover the moon
with your hand. Beyond that, the sun, and
then the shift to a vast swath of stars and
galaxies. All can be covered by your hand.

III

They are all true in relative size to 'you'
the conscious entity that holds out its
hand, as you are to them. Telescopes have
been conceived by mind to bring them
apparently closer. But this is an act of
evolving mind to see a part of itself that
is close enough to touch. It is as thinking
to bring that part closer. You move your
hand for a closer look. The cells comprising
your hand have such small individual
enlightened consciousness as to be 'almost'
oblivious to the greater consciousness. The
atoms comprising the cells have a lesser
consciousness still. It is the greater
consciousness, of which they all comprise,
that has moved them closer to their own
truth. The truth of the mind behind the
creative thought to will evolved eyes so
that they can gaze on them, which is a part
of itself. You can cover your hand with your
other hand. You can cover a sea of galaxies
with your hand, by reaching out, for they
are part of your hand, your greater self.

The enlit then illuminated that I wasn't
there. 'So, where am I?' It was a reasonable
question, I thought, after having such a
thought placed in my mind. They illuminated
that I was everywhere, as they too are
everywhere. 'But I can see the school, and
the kids playing. Some of them I'm going to
make friends with.' They illuminated that I
was experiencing through only a tiny part of
my true self. The rest of me lay behind a
universal door, which they would slowly open
for me, for the light beyond would be far
too bright and needed time to adjust to.
Before I could ask deeper the enlit vanished
with the school bell.

The spark tree is still there, everywhere,
as I write, as are all the other mercifully
saved trees around what had been the
schoolyard. It is now a gated enclosed car
park and games area. The 1908 physically
perceived school has gone. But its time
shadow remains. Space has recorded it.

V

For everything is recorded. The position of
every atom. It needs to be, for the
conscious possibility of recreation to
exist. Without recreation nothing actual
could exist. With sufficient conscious will
the school could be built again exactly as
it had existed physically, to physically
exist again from its space-mind conscious
entity recorded state. It simply needs the
will to move matter and create it. At some
point in time my conscious will to gain
access to the gated area shall succeed and I
will stand with that conscious will by that
spark tree. I shall reunite with the time
shadow that I left there so long ago now.
Time, relative to the gained experiences of
my conscious mind. The circle will complete,
as it has to be. I will thank the enlit. And
they will thank me. For my lit is also
theirs. My experience adds light to light.
By giving they give to themselves, as with
all giving. This is how spirit feeds.

VI

It is not possible to reason the true nature
of the enlit. They are beyond our reasoning
capabilities. It is as we would instruct an
ant to build a better nest, if we possessed
the powers of instruction. But we do not. We
can observe the ants with our greater powers
of observation, and our intellect to know
what they are doing, but we fail in the
power to communicate. Yet we are not the
enlit to the ants. They are to the ants as
they are to us. Their wisdom is to allow the
ants to be themselves, as Nature has
programmed them to be. Too much information
would crash their hard drive. Too much light
would destroy them. Destruction is not a
particle within the enlit thought concept.
We can destroy the ants, and do so. It is
the wisdom of the enlit to allow us this
pure thought failing. They allow creative
destruction. For positive thought has to
have negative thought to balance our astral
physical particle existence. We are not yet
ready for pure positive thought. The enlit
have wisdom to this. They are the balance.

VII

The wisest grain of sand is the one that
reaches to know there are other grains of
sand, and that they form a beach.

And by such knowing, all other grains of
sand have such knowledge, for being known of
is to know. The first grain of sand that
reaches through to a knowing awareness will
illuminate the entire beach. For that is why
the beach is there, in waiting.

You have reached to an awareness that has
the 'grain of sand you' to gaze upon a star.
And by so doing that star is aware of you,
for that is why it exists, and why you exist
to gaze. Both are wiser, for the gaining of
self wisdom is the essence of all substance.
All needs to know of itself.

You walk the beach, as you would walk the
sea of stars.

Enter

VIII

You have entered a concept enriched zone. To
continue you must open your imagination and
be willing to allow a shade of colour from
the enlit through the window of your inner
reality.

I trust you will continue, as do they. For
ALL continues and ALL returns to ALL source,
then to all continue again.

IX

There is a code of law that runs through the
entire structure of all conscious matter. It
illuminates as follows.

For any particle of matter to exist, the
reality of existence must exist, and for
that reality to exist the reality of
consciousness must exist. And for the
reality of consciousness to exist, mind must
exist. And for conscious mind to experience
reality existence of all matter, all matter
must be conscious with itself and of itself.
And in so doing it is itself all matter
seeking the experience of itself.

Consciousness = Matter Experiencing,
Squared. $C = ME^2$

> Truth =
> Temporal Reality Infinitely Perceived.

The Universe is not expanding.

ExiT   ConTinuE

X

You have entered at the sign of Capricorn,
the sea goat. Its fish tail swims in the
astral sea. Its front hoofs climb the rocks
of material substance. It knows both realms.
By allowing your wisdom to continue you have
accepted Capricorn and that reality has many
more colours. Everything is a doorway to its
greater self.

WordLight One.

To finish, one must first begin,
To conquer all, yet at once give in,
For all bright hopes, Time's trumpets blast
in vain,
What has happened once will happen again.

We meet each other, and yet we pass,
We learn, although there is no class,
Search for treasures that are there to lose,
So many things, we have will not to choose.

2nd Beginning.

Which itself needed a 1st beginning, as with
all beginnings. Beginnings are infinite.
Infinity is a constant beginning. To begin
is to have the beginnings to begin.

R.O.A.M.

Ramblings of a Matter-and-dimensions-man.

The Dream creates reality.

T. HuGotH.

XII

Contained within these enlit reached pages
are illuminations on reality.

An Explanation of the Unexplainable.

A definition of reality.

An expansion on Time

They are shone in twelve full conceptuals,
radiating as nine illuminations and three
consequence awakenings.

You stand at your own conceptual doorway.
From this point there is no exit.

WordLight Fourteen.

We set our sail across Time's immortal lake,
Unknowing of the journey we must make,
An innocence that never seems to fade,
Until the final crossing has been made.

With such wisdom, although begetting late,
Resolve all things within their natural
state,
Thus, with humbled eyes this majesty we
view,
At last in readiness, we face a journey new.

The enlit wish us to listen. We have shimmied into the 21st century to gatecrash a new millennium. We stand by a new pond. It seems now that our time line is ripe for a new splash of thinking. It is time to surf uww.com

Universe Wide Web. Centre of Mind.

We have been dangling our toes in the pool of knowledge and think we are ready to skinny-dip. But we have forgotten the fluffy towel of inspiration and our clothes are trampled with misunderstanding and then soiled by cow pats of ignorance on the grassy banks of innocence. We have been hasty and ill prepared. We need that towel, and the candy striped wind screen of understanding. We need the floral trunks of enlightenment and the goggles of wisdom. Let the crusts of thought provided within these pages be a new preparation. The golden egg of truth sizzling in the blackened pan of deceptions must not be allowed to burn.

It must not be allowed to go crisp around
the edge. Let us switch on the blender of
learning. We are ready to hear the music of
the stars, live

Conscious movement is Nature's investment in
space usage.

Reality is the difference between events
happening at the same time. Space is tight.

Reality is not what you think it is. This
now that you are experiencing could be
something else in its entirety.

Reality always rounds itself up and space
imports space from other space.

To explain and understand the nature of the
Universe requires a great deal of thought,
and a comfortable chair. We begin.

Illumination One.

The Universe is not expanding. This is an illusion. The perceived speed of light is also an illusion, as light is the slowest thing within the Universe. It is actually stationary and the perception of light is in fact a gravity pulse stimulating stationary photons to illuminate and transmit a perception of the origin source from photon to photon which our temporal conscious entity interprets as a transmitted 'light wave' image of that origin. We SEE something. It is as electricity passing through bulbs in a sequential pattern. From a distance there is an illusion of movement. Extremely close and they are mere flashes of energy at a stationary point. The quantum processor that is our brain evens events out to an acceptable level for our physical consciousness to accept.

We SEE events smoothly and in apparently real event time. We have built our physical instrumentation accordingly, which are fooling us. We see a light wave travelling out into space, of which there is no such thing anyway as everything is a form of solid. But it is Nature and its related spirits that has tuned our quantum processor brains to experience light at a certain quantum level. Nature has, for the moment, chosen, shall we say, 32 bit speed, as the full galactic telephone number bit speed of true reality experience would be too much for our still awakening minds to cope with. Evolution will expand our potential, if our base instincts remained controlled. But there is an aura of optimism around the conscious planet. It is delicate but relatively stable. The galactic concilium of the enlit deeply wish it remains so until a level is reached for communion. There influence cannot be too excessive. It has never been so. It can never be so. It is the gentle cultivation of a flower.

But if there is too much darkness within the soil, the crystal water will ultimately fail. Such signs as they have given and have physically left can be interpreted in many ways. Such was their design. Such was the way it was scribed. But at last we are suddenly truly accelerating into awareness. If the planet's positive aura can remain as such for a hundred or so more orbits of its given life star, then we shall have secured a tremendous welcoming.

This given digest is a light step closer.

The Universe is not expanding. It is in constant motion, but it is not expanding in the manner that our physical instruments would have us believe. If it were expanding, then what would it be expanding into? Expanding also assumes a singular point of origin. The Universe has never had a singular point of origin. It has never been anything else but the Universe, and never will be.

For it to be something else would be to deny
its own conscious existence and all life
forms within it.

As we are aware that we exist, then the
Universe is not in denial of itself. It has
therefore always been the Universe and never
any singular point event. For what comes
before a singular point event? There is much
talk regarding a large explosion of some
kind. This is a false trail created by the
perceived illusion of expansion via the
limits of physical instrumentation.

If you click your fingers then that is as a
singular point event. But what came before
the clicking of the fingers? It would be the
thought of clicking of the fingers. And what
came before the thought? It would be the
existence of an entity via which the thought
could transmit. And what came before the
entity? It would be the thought of creating
the entity.

This in itself would need an entity via which the thought of creating an entity could transmit.

The creation of the Universe from a singular point event is the only true impossibility. Nothing could ever produce something. The 'why' factor would have had to exist first and that in itself could not exist without a form of entity Universe. Your fingers would not click if you did not exist to think to click them. You are the Universe and your fingers act out a singular event i.e. the first combining of elements to create a star. Your fingers, the star, would not click, nor the star exist, without the Universe YOU, and you would not exist without a Universe to exist in, where stars are created for your perception. You, reading these words, could not possibly have come into existence from nothing. Neither could the Universe, because you are a part of it, and your existence reflects the greater existence of the Universe.

To create from nothing is to create nothing.
As you and the Universe are most certainly
something then something will always create
a something which is something. It created
the something that is you.

To say that the Universe began from with a
large explosion from nothing is to say that
you are still as nothing, and that you
cannot click your fingers.

The Universe is in a constant state of
conscious living creation and none creation.
Everywhere within the Universe is a point of
origin for an event of reality and creation
to happen. Your lemonade has the answers.

All manifestations within our perceived
levels of creation have an analogy that
Nature and its related spirits have provided
to aide our ability to understand. A tall
sparkling glass of lemonade on a summer's
day. The bubbles form within it and rise to
the surface. Universe bubbles.

Each bubble is as an event point within the lemonade Universe. Within the confines of the glass and the material constrictions of their dimensional framework the bubbles can only rise one way. The 'thought' behind the creation of the bubble is the chemical nature of the liquid, and the greater 'thoughts' behind the chemical nature of the liquid are the high conscious entities whom have mixed the chemicals. It is their thoughts of creation that have filtered down to create the bubbles within the glass of lemonade. True event 'bubbles' have an infinite level of directions to choose from.

There are lava lamps, which are upright and basically cylindrical. Some have tiny glitter particles drifting around with the currents of heat from the base lamp. Turn the casing into a sphere and you have an analogy of 3 dimensional movement within the Universe. Imagine the glitter particles having the ability to drift through into an adjacent sphere.

You have an analogy of inter dimensional movement. Imagine the lemonade bubbles drifting through the glass you are holding and floating around you. Now imagine them drifting back before the greater space becomes a danger to their glass of lemonade temporal form. The bubbles will have experienced your higher presence and the experience will enrich the lemonade.

Your walk along the beach enriches the grains of sand. You could, of course, stir the lemonade with your straw and oxygenate them with your presence in such a manner, should the bubbles have not reached the technical and mental capacity to transcend beyond the glass, which is of the greatest possibility. Your thought input, your 'finger click', will then create more bubbles. You are the something behind the something event. Take 'you' away and there would be no bubbles. It was you who poured the lemonade in the first place.

They exist because you exit, as the Universe
exists because it is constantly being
poured. And before the pouring there was the
thought of pouring. The Universe has always
existed, because thought has always existed.
The Universe is a glass of lemonade

Imagine the glitter particles, within the
spherical lava lamp representation of the
Universe, moving along a spherical course,
in line with the outer surface of the lamp.
Some glitter particles are moving faster due
to their larger size. Tiny ones will move
much more slowly, as they have less capacity
to be influenced by the heat from the bulb,
working through the suspension oil. However,
the light source is now no longer at the
base of the lamp, as it has no base, now
that it has become a pure sphere. The 'bulb'
is now itself a sphere, around the lava lamp
sphere. Thus, the glitter particles are
receiving heat, energy, will, consciousness,
from all directions. Yet they all move along
a general spherical trajectory.

This is due to the spherical confines of the suspension oil within the spherical lava lamp casing. With gratitude to Nature's wisdom to provide an analogy to answer every question, all answers are within eye touching distance, if the eye is opened wide enough. The lava lamp suspension oil acts as gravity, but in a much more simplistic way. But the analogy provides a starting point.

The glitter particles move along in a variable speed spherical trajectory powered by the bulb energized suspension oil – gravity. The energizing bulb also equates to the pourer of the lemonade. To an observer on any glitter particle the time scale of the spherical motion will be so great that other more immediately perceivable event phenomena will obscure the true motion, and will be on a much quicker time scale to add to the distortion. You can't see the stars under a street lamp.

The glitter particles move along in a
variable speed spherical trajectory powered
by the bulb energized suspension oil –
gravity. Given the confines of their
existence interface there will exist an
event point in time when they will arrive
back at there temporal origin point.

The bubble appears and moves upwards due to
its lemonade glass existence. Expand that
existence into the glitter lava light
Universe sphere and the bubble would
eventually arrive back at its arrival point.
Again, due to their varying size, some
glitter particles will arrive at origin
point quicker than others. This is logical,
as it would not be practical for the
Universe to have all its particles arriving
at source point at one time singularity.
That would be to have everything happening
at once, which it doesn't do. Even basic
corporeal sentient experience displays this
to us. Nothing around us happens all at
once. All at once is nonsense.

The Universe is most certainly not nonsense.
It contains the sense that gives us our life
sense of experience. The idea of a big bang
is an idea that our entire experience is as
an existant nonsense.

Everything takes a factor of time and true
thought to event happen. We are a true mind
mirror of the Universe, thus, if events
around us don't simply happen, the Universe
didn't simply happen. There are no mini big
bangs as you go shopping in ASDA and a WOW
moment as something comes into existence.
Would you want to 'happen' with a big messy
explosion? If the answer is no then it is a
tiny thought echo from the mind of the
Universe. The Universe didn't simply happen,
with a very undignified and inelegant big
bang next to the fruit and veg. It would
never allow itself to do. It has too much
high thought about itself. Big bangs are for
dizzy stars, not for Universes. And a true
Universe wouldn't allow oneself to go off in
a shopping trolley.

WordLight Eighteen.

I turn away to look you in the eye,

Laugh aloud with thoughts to make me cry,

I sing with sweet yet empty song,

And all my rights at once are wrong.

Bid turn back time, yet travel on,

Stand with crowds, still being as one,

Feel the needs of hope but its warmth grows

cold,

As young am I as Time is old,

So full with contradiction's curse,

And through loving, for the better, I am yet

worse.

The Universe is infinitely too sophisticated to allow itself to be created by a crude singularity big bang.

Just as the curvature of the Earth appears as a flat surface because we are so close to it, the curving trajectory of motion is totally lost due to its immensity. The tiny glitter particles within the Universe lava lamp perform mini spherical trajectories along the course of the major trajectories. Observers on a glitter particle would have a confusing vision, further confused by the density of gravity. If you look down from a small boat into the crystal waters of a shallow sea, the beautiful white sands of the sea bed, and the coral with its sea flowers, will appear to shift and ripple with distortion. If you had no acquired knowledge and limited depth of mind you might conclude that you were seeing the reality sea bed, and wonder why it rippled the way it did. Diving equipment would be beyond your comprehension, at first.

Your curiosity would enlit grow until you
discovered that a boxed piece of glass
placed in the sea gave a totally different
perception. Distortion vanishes and the
coral becomes open to its full experience
potential. Now you would spark the desire to
visit the coral. The density of the crystal
sea waters acts on a miniature scale in a
similar way to the infinitely vaster sea of
gravity. The gravity charged light photons
travelling through the crystal sea are
disrupted to create the rippling image we
perceive from the boat.

Light photons within the vastness of the
Universe are disrupted in the same way, but
on a much grander and in a much more
sophisticated way. Everything 'out there'
has a tiny mirror counterpart expression of
its reality 'down here'. The answers to
everything have always been scattered around
us, waiting. The crystal sea and the
Universe gravity sea are temporal echoes of
each other.

It is as an echo of sound is the sound itself a mirror of its original source. This was a source that had the thought to create the sound. The echo did not come from nothing.

The red shift experienced from distant galaxies, and thus used as a measure of their distance and speed of expansion, is a complete illusion, as the sea bed distortion is a complete illusion. A train whistle 'will' change in pitch as it passes and 'apparently' moves away from us. Sound frequencies are at a very low level on the Universal scale. They will be affected by a physically distorting atmosphere, and we also SEE the train moving away from us, acting as a confirmation of a perceived reality. This is how our quantum brain delivers information to our consciousness. The train is in fact returning to us, even though it appears to be moving away. Our brain allows us to think it is moving away, to avoid confusion.

Everything is in a state of returning. The primitive sound wavelength and its frequency will be stretched by temporal conditions, creating the experience of a change in pitch. And we also have the given experience of the temporal visual confirmation of the train growing thought perceptibly smaller. On a grander scale, the 'red shift' from the distant galaxy is caused by the thickness of the gravity sea. The further it is away, the more gravity sea the light pulse has to travel through.

The light pulses from our life-star are too brilliant for physical eyes when travelling through the sea of atmosphere from directly above. But, when travelling through a much thicker slice of atmosphere sea, as the planet turns on its axis, the wavelength frequency to our physical eyes is lowered and we experience a reddening. We have the joy of a beautiful sunset. Nature's gift to us, which is a tiny mirror of a much greater reality. The speed of light is not constant.

he speed of a pebble that you have mind

onsciously thrown into a crystal pond is

odified as soon as it hits the surface of

he pond, thus modifying the thought will

hat threw the pebble. A pulse of gravity

harged static photons is constantly

odified by the amount of gravity sea it has

o pass through.

magine the row of small closely arranged

ight bulbs extending a mile or so. You are

tanding a quarter of a mile from the centre

f the row. From this vantage point you can

xperience the full length of the bulb

isplay. The electricity (electricity that

as had a creative thought to generate)

luminates the bulbs (the light photons)

equentially, creating for your mind the

erceived illusion of a pulse travelling

ong the line of bulbs. You are observing

his from a totally linier, left to right,

one vertical one reality level. A true

hysical reality would include vertical.

A true series of light bulbs would be illuminated from a central originating generated power source. Detach your imagination from the electricity grid and imagine that each small bulb is powered individually, from a Universal source – gravity.

Each bulb now gets a tiny individual shot of power, creating the same sequential illusion of a pulse, from left to right, along the mile long trajectory. A collection of matter, a planet or a star or a galaxy, is a concentration of gravity. It is the same as a flock of birds. They have the collective concentration to flock.

Gravity is the power station behind light. To the right of you is a galaxy. To left of you is a planet. The galaxy is vast compared to the planet. The galaxy has a strong gravity presence, which powers light photos to advertise its existence to all sentient entities.

But the vast depth of crystal sea gravity
will dilute and distort the advertisement.
You, the observer, would see the sequential
pulse slow down, as it ploughs against the
myriad of other pulses which are cross
directional 'cross dimensional' to your
event spectrum and are also invisible to
you. For you, this is an atom point one
dimensional sliver of event time. To see the
greater picture of events, turn away from
the mile long line of sequential illuminated
bulbs and look up into a clear night sky.
That is a view closer to the reality.

WordLight Twelve.

If we should stumble, shed no idle tear,

Survive and learn, but not in fear,

All is not wasted. Survive and test the lie,

Of all things ending.

If so, then why, this way, pass by.

This cannot be a simple fleeting thing,

A rough design in need of finishing,

There is too much polish, and all the pieces

fit,

And here we are, the greater part of it.

The red shift is an indicator of distance, i.e. depth of gravity, not of expanding speed. The Universe is in constant motion, but not expanding from a central point. Various points will drift away from each other, as dose the glitter within the lava lamp. But these are only localized events within the lamp. It has - at last - been physically observed that the further the vision increases, fully formed galaxies are still physically visually experienced.

This is a strange paradox, as the expansion concept would have it that out is also time back. Creation did not just happen. It is always happening. The further you look out and you will still see it happening. But the further you look out, the further along the curve of the Universe lava lamp sphere you will observe.

Unlike our planet existence constraints, we are not, at Universe level, stationary to the horizon of the curve.

We are constantly moving towards it, due to
the lava lamp oil of gravity. Should
technology be allowed to expand, as it is
guided to do so, but not compelled to do so,
and our base ancestral motivations
controlled, devices may yet transpire that
would look far enough out to actually see
ourselves. And we would see ourselves
stationary, because we would have completed
to the curve. We would have then returned
back to source, in a visual and mental
sense.

We would still be moving, 'expanding' but
the movement 'expansion' would be parallel
to ourselves. And it would be observed for
we had the mind to do so. Nature provides
every example. A salmon swims upstream, back
to where it was spawned. A particle curves
around the Universe, back to where it was
first thought of as a particle. Back to
where the salmon was first thought of, by a
breeding pair.

The time span is infinitely greater, the depth of gravity crystal sea water is infinitely greater, but the event will still happen in actuality.

There is only so much space within the Universe lava lamp sphere. Some particles will be quicker, others will take ages, but all will eventually reach origin point, then to regain energy and continue again. If it gets a big enough kick, it may well return back to source quicker next time around. The train whistle changes in tone as the train moves into the distance, and becomes temporally physically visually smaller.

Universe events are observed with physical instrumentation as accelerating away from source at approaching light speed. And we are, in theory, accelerating away likewise. The train does get smaller, as we visually experience it at the speed of, for example, speed of light, 100mph maximum experience level.

But are you accelerating along the platform
at 100mph for the privilege of experiencing
the train accelerating away from you at
equal speed. If this was reality, nothing
would see each other. We would all be at the
speed of light.

The galaxy, strangely, remains the same
visual size. The theory is that we see it as
it was, because of the time it takes light
to reach us. But if 'as it was' is still
that distance away, then why is 'as it was'
not whizzing away from us at near so called
light speed? If the Universe is expanding,
at close to so called light speed, at the
outer limits of our visual capacity, then
why can we still observe events? A point
would be reached where the light would not
escape because its source was travelling at
the same speed. Light would have to travel
faster than itself.

If looking further out also means looking
back in time, because of the span of time it
has taken the light to reach us, we should
have visually experience younger events such
as galaxies forming and eventually nothing
at all. But we visually experience fully
formed galaxies, which must have taken time
to evolve, yet they are at an alleged point
in time when they shouldn't be as they are,
because there wasn't enough time to do it.

Our minds are slowly being programmed for a
hard drive update. Questions are being asked
regarding the 'big bang'. Those questions
are the first u.b. sockets installed.

WordLight Thirteen.

Yes, we stand and dream and think it all in
vain,
In hope, all things done turn this way
again,
But if each cloud does truly fade away,
Why give us breath to marvel in each day.

What force wastes time assembling such a
plan,
For no reason, other than it can,
Such idle thought does injustice to Its wit,
For if all things end in nothing, so too
must It.

Illumination Two

Light is a pulse of gravity charged photons,

creating the mind experience of a light

wave. The speed of the 'wave' is governed by

the reaction capacity of the photons. Their

maximum capacity has been recorded with

physical instrumentation at 186,000 miles

per second, which is incredibly slow on a

Universe scale. But it is gravity that is

performing the magic we perceive as light.

The mile long line of bulbs can only cope

with a shot of electricity up to a certain

time frequency. Beyond that and their

filaments would break due to being turned on

and off so quickly. The wave illusion would

be measured at 186.000 miles per second. As

there is nothing visually perceived to be

travelling faster, and as we are under the

assumption that it is an actual travelling

particle and not an illuminated static one,

a light bulb on the line, we can construct

theories and thought platforms.

But these only fit temporally perceived events. The blueprint is upside down. Photons are static and omni present. They wait at their dimensional level, ready to be charged by gravity and transmit a tiny fraction of an objects physical reality from one physical space point to another, so that it can be perceived by a sentient entity. Gravity would not do this just for the fun of it.

Sentient entities need to exist to have the experience of the gravity action as light. Why would gravity bother to stimulate photons to create a light wave experience if there were no conscious minds to perceive it, other than itself? Why would anyone build the mile long line of light bulbs if there would be nobody to watch the pulse, other than the builder? There is always thought before a physical event, with the secondary thought that it will be thought experienced by other thoughts of equal wanting to experience the event.

Such is the nature of existence. Thoughts
need other thoughts. Events need other
events, so that events can actually event,
so that more thoughts can think of more
actual events.

Can you possibly think, have thoughts of,
nothing? It is the only true impossibility.
The thoughts themselves are something.
Thoughts of nothing come from something.
Nothing comes from nothing. The Universe did
not come from nothing. If it did, you
wouldn't be able to think about it coming
from nothing.

The light photons are static and omni
present. They are gravity charged into a
light wave perception. They are also charged
with a time perception. Gravity is the core
of all material existence experience.
Nothing physical would exist without it.
Time would not exist without it.

Physical reality time perception is
necessary for enlightened thought
transmission to create a consciousness that
is capable of observing itself.

Look into a mirror and you are the Universe
observing itself, because it took thought to
create the mirror. And you are also taking
the thought time to look at yourself. The
thought time that created the mirror also
created the now time to look into it,
otherwise it would not have created the
mirror. Everything is created for something.
A chair is created to sit on. The Universe
is the chair for thoughts to sit on. The
Universe is sitting on itself. The chair did
not arrive with a big bang. It would have
been in all the newspapers. The chair is a
physical entity reflection of the thought
that created it. And the thought that
created it did not arrive with a big bang.
As you read this and have thoughts, are they
arriving with a big bang?

It would be wonderful for the manufactures
of headache tablets, which they themselves
would be using, having big band wonderful
thoughts of all the tablets they are
selling. The chair creating thought came
from a much greater entity than the chair,
otherwise the chair would not exist. And the
chair creator, with chair creating thoughts,
is itself a wider reflection of the Universe
entity. The chair is an entity in its own
right, because it is a distilled reflection
entity of the full Universe entity, created
by a distillery of creative thought.

Perception =
Permanently Interactive Sentient
Spatially Expanding Dimensions.

As you look into a mirror, you observe your
own thought creation, distilled through the
Universe, because your thoughts reflect
through the Universe, and they reflect back
to create you.

As you look into the mirror, the Universe
is, shall we say, checking, through your
mind, which it created, that the reflection
justifies the reflection. The Universe needs
to see why it is allowing such 'looking at
itself' thoughts. Through you at the mirror,
and with all thoughts, through every level
of consciousness, down to a single, human,
animal or plant form conscious living cell.
And then further down to mineral and base
rock. The Universe observes its own sentient
creation, its own thought to create, and
then experiences all existence, which is
itself. The Universe is extremely vain.

The Universe could not possibly be expanding
as is physically observed by illusion shaded
mind created physical instruments. As
gleaned, it would have to be expanding into
something, a something that is beyond
itself. It would be expanding into a thought
something that is nothing, as it has yet to
be thought of. As shone, the Universe is all
things, other than nothing.

As shone, it is impossible to think of nothing. Thus, the Universe cannot think itself into expanding into something that is impossible to exist. The Universe is in true constant motion and creation, through the sea of gravity, but it is not expanding. Would you want to expand your size infinitely? Of course not. And that thought is a tiny reflection of the Universe thought, because you are dream part of the Universe. You are a tiny thought expression of the Universe. You are created from it, and it is created from you.

A single cell within your body is quite happy doing what its infinitesimally tiny portion of sentient thought allows it to do. You are its Universe. You are the collective of all those thought particles. But on its own, it cannot possible conceive that you enjoy going out to the theatre, whatever, and for a drink with other Universe friends. It has its own tiny positive thoughts to do its job and keep its Universe stable.

But everything has opposites. All mirrors must reflect the image presented to them. That is their job, to reflect photons. That is why the thought existed to create a mirror. That is why a beautiful still pond of clear liquid thought existed so that high sentient mind could gaze in and observe themselves, and thus have thoughts to improve on that reflection – and thus fashion metal and then glass, thought manipulating matter, to create what we all know and love as a mirror.

You are the Universe to the thought happy positive cell. But, as everything has its opposite, its reflection, there are negative thought cells. Their distilled thought is to destroy their own Universe, and thus, themselves. They are insane, as with all negativity. It has only none creation to offer. Creation is the essence of existence. None creation is the essence of insanity as it goes against its very reason for temporal existence.

But insane cells have to exist to comply with the negative & positive Universal rule. To be reality infinitely sane, infinitely perfect, would have to be as having reality nothing to judge itself by, and would thus have reality go reality infinitely insane. This is the reaching danger of creating a perfect mirror.

This enlit given digest takes us to where we are fully Mind Aware Dimensionally.

WordLight Two.

To gain a step, yet lose a mile,

Is through the gaining still worthwhile,

For if we never once did gain,

We'd be before ourselves again.

To swim the sea and never sink,

To stare out the sky and never blink,

Our limits make us what we are,

And with these in toe, still we travel far.

Freedom is a captured state,

Willful, that it comes too late,

For all things lost for little gains,

Such freedom, blessed, to choose our chains.

Space is the visible shadow of event time.
If there was no motion there would be no
space and no time. The Universe would be the
ultimate solid that it is always trying not
to be. Through motion it is at once solid
and at once space. And yet space is an
abstraction. The act of a motion generated
event creates the very space for the event
to happen.

You stand at the far end of a large room. At
the opposite end there is another wall and a
window through which the sun is shining.
Your mind perceives the far wall and the
sunlight through your given senses. There is
also a sense of space in-between. There is a
perceived distance. Now you have a conscious
will to move towards the wall and look
through the window. You become a motion
generated event. Now the space is made
available to you and you are allowed to move
towards the wall. Expand imagination. If you
rolled a ball towards the wall then space is
being created for that ball.

You perceive this as the ball moving into a perceived distance. The space is being created because the ball was moved by a conscious will. The ball could not move by its own will and so no space would be created for it. The ball would be a perfect solid. But it is a ball prevented from becoming such a state because the conscious will exits that has the power to move it. 'Nothing' could not move it.

Nothing =

None-directional Universally Transported Space.

The Universe, as we perceive, began because it wanted to begin, and to end. It wanted to begin and end at the same time. It is constantly beginning and ending, and dream experiencing. It has dreamed itself into continual existence, knowing that it was always a dream of continual existence. This is good news.

Being part of the Universe we can crack open
a can of beer knowing that we will have some
form of continual existence. Cheers to the
Universe.

Matter reinvents itself as it always has
done and always will. It is asked what came
before this so called 'big bang'. The answer
is quite simply all around you. It was the
'big everything'. Everything came before
because everything was waiting to happen, as
everything is waiting now. And everything
could not possibly be nothing. There is no
such thing as nothing. There is always a
something, waiting to become something else.
In this case the transition is perceived as
this stupid 'big bang'. But we perceive
event happening matter with instruments
consisting of such matter. All matter is the
outward expression of the universal pure
thought that is the true reality. Its dreams
are our reality.

Dreams =

Descriptive Omni Telepathic Trans-
dimensional Interface Experience.

The ball could not move without our dream to

make it move. Through that dream we give the

ball energy to move. We give it the thought

life to move. And we create the time for it

to move which in turn creates the space for

it to move. And it is a thought space that

is thought waiting to be created. We can

perceive the distance to the far wall and

the window through which the sunlight

shines, yet we have not physically moved to

the wall. That experience awaits us. That

thought space awaits us. The time awaits us.

The thought time. The event will happen.

But first, there is the dream of the event,

the 'looking across the room'. The ball

transmits a particle of that dream. The ball

is moving through our dream for it to move

and transmit a fraction of that dream across

a dream generated expanse space and time.

Our existence is communicated to the wall, which is thought waiting, with the window through which the sunlight shines. Before the existence of everything 'we' see there is an existence for someone, some event, something else to see events happen. But the seers, being ourselves, are only the base instruments through which the essence of Universal thought wishes to experience 'to dream' events. If we physically wish to see a distant star we use a physically engineered telescope. The dream behind the telescope does the same, on an infinitely more subtle level. Through you it dreams of being a rolling ball.

Strike a well gripped crystal glass and it will sound dull. Release and it will sing.

WordLight Three.

Whatever distance falls between,

Still so close, the great unseen,

Shaded quiet by your side,

Through time and space, no place to hide.

But do not let its presence scare,

For you alone it is there,

Soft and secret, your gift to touch,

If the mind so moves, then move it much.

Everything is:
Relative Active Mind Building Logical
Existence.

As I ramble, so does the Universe. If the
Universe is rambling then we all are.

Illumination Three

Gravity is a particle. It is the ultimate
particle. It exists at the true edge of
material consciously physical existence, as
designed by Nature. It is so small for the
exact reason that its existence is to hold
the particles of known atoms together. It is
as an atom within a drop of water thrown
into the sea. Gravity is the drop of water
forming the drops of water that forms the
sea between all atoms. Gravity is the
supreme particle. It is so tiny, but it is a
chess piece capable of acquiring every
movement factor from all other pieces. It is
every piece on the board and can make a move
according to the situation – according to
the will of the player. It is that fantastic
mythical chess piece that can suddenly
become anything and save the situation,
according to the will of the player, whose
will brings that magical quality into play.

But the gravity particle is a much smarter chess piece. It is a consequence particle. It can adopt the moves of all the other pieces – but only when 'it' chooses the appropriate moment within the game. It is a chess piece that moves itself and changes its piece mode to meet the challenges of the game – the Universe. The players can only watch in awe as the chess board expands around them and they find themselves within the game.

The gravity particle is a magical any piece chess piece because it holds everything together by tying matter through its own gravity induced timeless dimension. But here we have a paradox. 'Gravity' is the 'result' of this supreme particle's activities. The particle itself is not gravity. It is a particle designed exclusively to create what has been termed gravity, which is needed to hold conventional matter together for it to do what conventional matter is designed to do, create stars, flowers etc.

Without this supreme particle nothing would happen, and nothing being the only true impossibility then things will continue happening. The supreme particle will continue doing its job, because it cannot nothing itself away, and never can do. There will always be a chess move that it can make. Nothing = Never. There never can be nothing. Hence, there could never be nothing before the much hyped big bang. There had to be something, and that something was the Universe.

Tomorrow never comes. Tomorrow. This is just a mind created word relating to active mind anticipations of future events that cease to become 'tomorrow' once the time and mind horizon has been passed. 'Nothing' is a similar empty mind created word that helps us relate to a thought space between events. 'Nothing is happening'. This is a pure impossibility. You are thinking 'nothing is happening'.

That thought is within itself something.
Nothing is the negative thought reflection
of something.

The supreme particle creates our experience
of gravity. It is the result of the shift in
matter caused by the particle's multi-
directional travel through co-existing
dimensions. A large cruise ship passes
relatively close to a motor yacht. The
passengers on board the much smaller vessel
will experience a rise and fall as the great
ship disturbs the sea around it due to its
powerful motion. Now, imagine that the
cruise ship is transformed into a pea sized
object, yet retaining the mass and the power
to disturb the sea with its motion. The
motor yacht people will still experience the
rise and fall, but they will see no visible
cause. 'Hey, there's that thing called a bow
wave again. And another one'. Extend your
imagination further. A cruise pea is now
travelling port side of the motor yacht.

One is also travelling starboard side, one
to the bow and one to the stern. The four
quantum cruise peas will create a multiple
bow wave. Our motor yachties will experience
a slightly more enhanced rise and fall. It
will have the added delights of side to
side, some tilting and rolling, with some
merry pitch and yawl to create the full
quantum Dolby surround-sense experience.
'Hey, there's that thing called a multiple
bow wave again. And another one. Shit, the
six packs are sliding off'.

At our ultra sophisticated level we do not
constantly yell, 'Hey there's that thing
called gravity'. The experience of gravity
is infinitely constant at terrestrial
levels. The 'bow wave' effects are beyond
light speed and our quantum brain gears
experience down for us to experience gravity
at a smooth constant terrestrial rate. The
influence of a variety of substances causes
the brain to become quantum unstable.

This results in the many humorously relatively harmless classic smashed experiences, or shading negative to raw aggression and then beyond to the mind break of quantum dimensional reference, resulting in tragic physical circumstance.

These substances:

Destroy Relative Universal Gravity Space.

Ultimate particle is a slight prosaic. Gravoid or similar grav sounding words are a little obvious. It richly deserves a mind enlightened thought name. It is at the horizon of temporal matter. 'The world of spirit lies beyond the veil', a quote from many with the gift of trans-dimensional resonance. But what structures the veil? The zint. The ultimate particle is the last dream of material substance. Its trans-dimensional activity creates the experience we call gravity. Distance is meaningless, as with the time taken.

The particle is everywhere at the same event moment. The sea sick motor yacht crew don't realise that the quantum cruise pea to port, the one to starboard, bow and stern are actually same quantum cruise pea, merrily flitting through dimensions i.e. port, starboard, bow and stern, which are all relative to the motor yacht.

We are the motor yacht, and our port, starboard, bow and stern are at infinite levels 'astral pond surfaces' around us. The ultimate particle sends its bow wave through them all. We experience gravity, due to the Universal directional influence of the zint particle.

The Universe is not expanding. This is an illusion caused by the density of the zint particles. The red shift is indeed a true indication of distance, but not the speed of the object moving away from us.

The perceived red shift is the result of the speed of light induced photons being modified by the zint particle gravity creating sea. It is the same as looking at the rocks, sand bed and fishes through a conventional crystal sea. The rocks and sea life appear to distort. The liquid sea is acting as a gravity field, on an infinitely smaller scale. The zint particles also help to keep 'ordinary' matter in a relatively stable condition by energizing conventional particles.

If a C particle needs a positive boost the zint becomes aware of the need and flies to supply it. The zint will then fly to a C particle in need of a negative boost, which could be trillions of miles away in our space reference. The time taken for such activity is instantaneous because the zint is beyond our time reference. Its actions create our time reference. As far as it is concerned, it creates its own time for all event action.

This, to our temporal perception, would be instantaneous. It is as a car travelling along a motorway. A car with a super will and mind of its own. The car doesn't like the distance it has to travel to reach its desired destination, so it compresses the motorway accordingly. The zint goes one better. It creates the motorway of our space and time experience, and being the creator, it can un-create the motorway for its own creative use.

Thus, one zint particle is nursing zillions of C particles, and all zints are doing the same, everywhere, at the same moment. It is this movement, this energy transfer of the ultimate zint particle, capable of becoming either positive or negative, that we experience as the force known as gravity. The zints are so small as to be almost none matter existent, yet they hold all C matter together. They are so small that they are on the edge of imagination which is the horizon between matter and thought.

In fact, they do indeed break through into the realm of thought, as a dolphin breaks through from its sea realm into our air realm, falling back again with gained and thought enriched experience.

Zints are truly magical. They also hold C matter together across our concept of time. This is possible because their motion creates what we experience as gravity, which in turn creates what we experience as time. The super-will car not only shrinks the motorway to get to its destination quicker, by doing so it compresses the time it gives itself to reach that destination. The zint particle creates its own zint time, which magnifies into our experience of time. At their level and speed the 'past' and the 'future' are a single curve of motorway. A roundabout motorway, which they can cut across to any point of C matter. They also 'remember' where matter was and 'remember' or predict where it will be.

This is possible because they know where a C particle was in the 'past' and can dream its trajectory through C space and C time. They give the C particle some energy and then rush over the central reservation of the motorway roundabout and catch up with it at its 'future' state. This in turn will be a 'past' for another 'future' state and so on into infinity, as two mirrors facing each other. The zint particle memory is an essence of the event which it absorbs and thereby creates a sense of knowing within the Universe.

The zint particle carries an essence memory of events across the space and time that it creates for itself. It is as the nerves in your body carrying an essence memory of your hand. You move your hand from left to right. Your mind knows your hand existed to the left and it knows that your hand will exist to the right, even though it hasn't reached the right space point yet.

Move from right to left and your hand will have a hand memory of the right event, through the greater capacity your full mind.

Zints constantly flow through us creating the gravity we experience. Their essence memory also flows through us, channelling especially through our minds. If someone's mind becomes shiny and in tune with their resonance then precognition or past affiliation is experienced. Your mind has then temporally slipped onto the central reservation of the motorway roundabout. It can see all the cars going around. Where they have been and where they are going. Some have the gift to will themselves to that point, allowing the essence memory of the zints to resonate freely. They are the profits and the seers. Zints are so dense that space is actually totally solid, but not at the same time, because they move through time as well as space. If they didn't, if they were all here at the same time, space would be a totally solid sphere.

Their magic sparks the light photons creating perceived light waves, and with these you can look out far enough and look back at yourself. In so doing you would be sharing some of the zint's magic and travelling around the motorway roundabout instantly. The zint multi-dimensional motion creates the sea for all C matter to float on. Take a large frying pan. Fill the pan with water. Sprinkle some flour on the water. Turn the pan slowly with the handle. The particles of flour will begin to swirl into spiral patterns. This is how galaxies are created and how they retain their shape. The water in the pan is zint created gravity on which C matter 'the flour particles' floats. Now, imagine the water surface tilted through every possible angle yet retaining its quality to float flour particles. You have an analogy of matter held together by zint particle motion created gravity.

The Universe is a frying pan.

WordLight Four.

To dance and never trip,

To skate and never slip,

To eat good fare, but never ask for wine,

What madness betrays a world so fine.

To turn around and face before,

The past is mirrored as once more,

And looking closely we shall see,

All the things we once should be.

To crawl, to walk and then to run,

Thinking our journey has just begun,

But compared to the wisdom of it all,

Our running is still, and ever will be, but

as a crawl.

Materiality =

Modulated Accelerating Dimensional
Differentiation Externalizing Reality.

AWAKENING TWO.

The enlit reached to me again a week or so
later. I was playing alone by a woodland
stream. Suddenly, the trees and the crystal
water seemed to have an extra glow. The sky
became a richer blue. There was a feeling of
happy warm electricity in the air and my
nerve endings were tingling. I knew they
were with me.

This time I said hello. There were flashes
of light within the stream and I knew that
it was their return hello. They speak
through ideas and it was my idea to answer
my question regarding where they were from.

I touched a tree. I touched the grass. I

placed my hand into the crystal stream.

I held my hand over the sun. Their ideas

told me that they were as close as

everything around me, but I could not see

them, for they stood behind everything

around me. The flowers by the stream were

the opposites of the flowers by their own

stream. Everything seemed a little brighter

as they were allowing the reflections of

both realities to touch slightly. You look

into a mirror. The surface of the glass

bends slightly towards you and the mirror

you. It is a reflection, but it wouldn't

exist without you to create it.

My ideas told them that I understood and my

ideas were pleased for me and told me to be

careful not to fall into the stream as my

eyes were too open. I asked what it was that

they had reached to tell me. My ideas were

simple. That I was having such ideas, and

that I understood them. That I would not

have understood without them reaching.

I watched their proof in the dancing light
on the stream. It was a light that I had
seen before and yet not seen before. This
time I knew the light could also see me with
its own ideas. I fell into the light.

My face submerged, and I saw.

Everything is seeing. Everything is
experiencing. All things see all things.
Everything is an experience event. All
events are ideas to have the event. Ideas
are flowers of mind. The Universe is the
ultimate garden.

With this I stood and walked out of the
stream, the light dripping from me. I was
soaked in it. But this was the wisdom of the
enlit. Now I was ready for their next
reaching, and beyond. They had stepped back
from the mirror. I knew, for the brightness
of the reality around me had dulled. But it
was not as dull as before. There was still a
slight afterglow.

Look around. At this point you might just
see it. We have touched in thought reality
through these pages. The enlit reach in many
ways.

WordLight Six.

Is Time so clever, as to go on for ever,
I enlit not, for where - when did it start,
To invent itself, from oblivion's shelf,
Would be an act of great mental art.

What came before Time? A question sublime,
With no time there could not be a when.
But there was and Time ran, at some point in
time, it began,
And what begins must end, now, and then.

Illumination Four

To begin, in the beginning, there was no
beginning. Beginnings and endings are the
same. Beginnings are simply reverse endings
as endings are reverse beginnings. They are
mirror images of the same event within the
continuum of thought. This is the core
principle of multi-quantum physics. None
particle matter does not begin. IT can be
the beginning of a particle event but IT
itself has always been there to create
beginnings. Once beginnings are created so
too is the ending of the event, remaining to
be reached through the course of time. This
'time' depends on how much thought has gone
into it, i.e. into the event in the first
place. Thought is the none particle reality
on which substance hangs. It is the washing
line for all that we see and experience. It
is our whiter than white reality.

Our telescopes and electronics are all
subject to the fundamental laws of thought.
They can only tell us a fraction of the
picture, because, like everything else, they
are part of it. We perceive a picture within
a magazine. A closer look reveals the pixels
which make up the picture. Closer still and
the molecules are revealed. Beyond this
would be the atoms. These are unaware that
they belong to a magazine photograph on
which we gaze. This would be their truth.
But the actual truth would be for us to
discover. And it would be in our power to do
so. The atoms might just understand that
they belong to a photograph, if they evolved
sufficient individual thought. We who look
at the photograph could discover who took
it, where it was taken and those whom are
represented. We have the greater thought to
do so. Yet we are a collection of atoms.
Thus, indirectly, they learn the true truth,
though our infinitely greater consciousness.

An atom within the ball learns the greater
truth that it can move across the room as a
sub sentient life entity 'ball'. It moves
with the will influence of an infinitely
greater consciousness, which is itself
comprised of atoms that mirror the atom's
true self.

The 'big bang', must have been truly nerve
racking to anyone living close by. The
ultimate noisy neighbour, if noise had
existed. Anyway, it would certainly warrant
bringing in the washing. It would have been
a truly fundamentally gigantic event, by the
limits of our experience, taking place many
billions of years in the past. But then,
there is no such thing as the past.

WordLight Five.

Time dances with us fast and short,

And stars light up our play,

As dreams, in mischief, through our minds

cavort,

With splendid disarray.

Though all our quickening steps be fleet,

We try to capture more,

Transporting every moment sweet,

To furnish well our lonely shore.

And though the tides of grave mishap,

May foam raw upon our beach,

The greater strength must never sap,

The will for stars laid in our reach.

For all of Time's undoing woes,

On faith, our hopes we build,

Then soon the day our wisdom grows,

And all star lost fears are stilled.

Illumination Five.

Time is an invention placed in our minds by
gravity. It is a means by which mind can
separate universally consecutive events.
Gravity thought creates time to form space
for it to think in. The same applies to the
future. As with, what is known as the past,
there is no such thing as the future, merely
events waiting to happen 'across' time.
Telescopes etc. are said to look back in
time. They actually look across the gravity
sea of time to events that were always
happening. The expansion of the Universe is
an illusion that goes deeper than the
gravity sea on which it floats. The further
away an object is, and the faster it appears
to travel, are the visible effects of its
closeness to its starting point. Movement is
a wave of Universe thought.

The greater the spectral shift tells us the
depth of thought, light being its shadow.
The Universe is, in fact, constantly
returning.

With gratitude to the enlit. Time is pure
consciousness. You think in time to think.

There is, on your time-line, an hypothesis
that one could travel back and change events
which would delete your history. You would
return to find your family etc. none
existent, thus erasing yourself. This is a
none reality in the plural corrective. It is
impossible to change events relating to your
history, or your future. If so then you
would not be in a position to time travel.
The fact that you have time travelled means
that all events on your time line have
remained constant, creating you and your
ability to time travel. Events in your
history can be changed, but this would set
up a new time line. You would return to the
history you had always known.

The new time line would continue in secular.

Great injustice could be undone, but alas,

on your time line the victims would still

suffer. The Universe is capable of

processing an infinite number of time lines.

It can be compared to a fractal pattern of

never ending complexity, but to add to them

with unnecessary vanities is a misuse of

ability.

This usually applies to meeting oneself.

There is a golden rule. Never converse with

yourself in the past, or the future. To do

so would set up a co-dimensional trans-time

quintilax. A super parallax along many

planes, all curving to eventually meet and

cancel each other into an original time-

line. It is quite possible to return or go

forward within your life span. The object

would be to observe only. In fact,

observation is the only logical procedure.

But you are existing twice I hear you ask.

Material atoms are changing constantly.

Mind, spirit and soul are none-time and can exist along side each other along different material time lines. But a meeting with oneself is somewhat pointless anyway.

To return to the past, meet your past self and change some painful incident in your history would simply set up a new time line for the past you. The 'you' from the future, as stated, would return to the same time line were events had remained the same. The only difference being that you had returned with new memories. The past you would have memories of their future self which would thus change their time line. Your comfort would be in knowing that the you on the new past time line would not suffer the painful accident. Of course, the memories that you have returned with would influence you and thus set up a time line for yourself which would not have existed had you not travelled into the past.

The same applies to future travelling. If
you saw yourself having some rather painful
accident and then went back slightly to warn
yourself, this would not prevent the
accident from happening on the original time
line. On that time line you didn't know you
had time travelled. You were un-warned. At
the very instance of intervention a new time
line is created. Upon returning to the
originating time point you would have
memories. You would have warned yourself
twice. The painful accident would still
occur, on the time line where no time
travelling or warning had taken place. But
any forward travel would involve memories so
you would know anyway and wait for the past
you to arrive and observe, or meet. Such
trans-time contra fluxes merely clutter up
the continuum with unnecessary and illegal
quintilax time-lines. They can also prove
hazardous to mental well being.

You would be in danger of becoming just a
tiny bit multi-astral disorientated.

Time travelling in either direction would not change events relating to 'you' the traveller's past or future. You would return to your origin point (with memories) along the same time line. The zint structure which created you and your ability to time travel belongs to the time line you have travelled. It has to return along the same path, or track. The new track that you have created due to your intervention will continue 'alongside'. Everything exists 'alongside' a reflection of itself. This is how everything knows about itself. Thought is a mirror.

Thought created a physical mirror by which to see itself at a perceived physical level. By seeing itself it created the time to see itself, and the thought-space in which the thought-time could exist.

Returning to the expansion on quantum time modulation, amplification and cohesion.

The Universe is a burnt frying pan.

The ability to jump tracks and witness how events have transpired requires very advanced zint amplification, which I am currently working to perfect. Zints, being almost none material matter, can shift between time lanes in the same way they can shift back and forth though time along a fixed space-time frame, i.e. the track you have travelled, or the motorway. To jump tracks you first need to establish the new quanti-space zint resonance that you have created. This is the multi-space signature of the new time track. Upon returning to your origin point observation is then required to signal out, with suitably amplified zints, those zints with the appropriate quanti-space resonance. Once established then the same poly-quantum dynamics apply as in standard time travel. Instead of travelling along a quantum time line, the motion is across. Cross time line observation is then achieved. The technology has actually already been dreamed.

Time and warp travel are at the horizon of
possibility, but the application is still
incredibly difficult. Warp travel is dream
possible. You have to be very Mind And
Dimensional. Leave it with me.

WordLight Seven.

If I started on a backwards train,
Would I, in consequence, begin my start
again,
How would I know my journeys end,
If back were front and straightness were a
bend.

Is up coming down in reverse,
Are all directions at once transverse,
For if I am standing still, yet moving fast,
Have I, at some point, myself past.

Perception =

Spatially Holographic Inter-
dimensional Thought.

Illumination Six.

Movement is always at the perceived sentient
experienced optional viewpoint.

All time travel involves space travel. The
ship would move through time but remain at
the same point in space. As space is in
constant motion the craft would instantly be
left behind by the planet, the solar system
and the galaxy. This is a simple fact that
bedevils the classic 1960s film, THE TIME
MACHINE. No explanation is offered as to why
the time machine stays at the same physical
point on the planet.

We must assume that Rod Taylor's character knew about Sub-Matter Astral Relative Thought.

All formal C matter elements have their own temporal resonance, their own dream frequency, which sings throughout all existent levels, just as birds sing in the woodland. It is the element's private multi dimensional E-mail address. It is their S.M.A.R.T. side.

To know this smart side is to know where each element can be found, and for other elements to recognise and harmonise with. The components a tree, although each having their own vibratory signature, their own smart side, will harmonize to form the sub-residual atomic resonance for the whole tree, because they 'know' they are part of a tree. Tree A. This resonance will, in turn, be similar to other trees, the tree world's sub-atomic recognition song.

But all will still retain an individuality
that will forever identify Tree A as Tree A.

All opera singers sing opera but we can tell
which one is Pavarotti. The principle works
upwards and downwards, left and right. A
pebble, a blade of grass, a flower, a tree,
an insect, an animal, MAN. All are component
parts of an entire conscious planet. We eat
fruit that grew in the planet's soil.

All component parts have their smart side,
which harmonize to form the planet's full
smart side. All formal matter is alive. It
is LIFE. A chunk of rock is LIFE. It doesn't
do much because its consciousness is so low,
but it has component factors that exist in
you, looking at the rock and thinking it is
just a rock. You think you are a smart
human. The rock, in its own rock way, thinks
that it is a smart rock. Everything is LIFE
at varying degrees of consciousness. The
entire planet is conscious, because we are
conscious.

But with it having the greater mass, it has

the greater consciousness, which it informs

us of in a variety of ways. Hence, crop

circles, the true ones.

If you are a planet, with humanoid matter

evolving on you, how do you communicate

without blowing their rapidly developing

minds? You control matter smart side with

your own advanced mind, which is everything,

to form patterns in plant matter smart side

with the hope they get the message.

Unfortunately, human spirit mischief has

confused the message.

Advanced mental attuning can harmonize with

any formal objects smart side. Controlled

matter smart side explains mysteries such as

the pyramids. The master builders of these,

and other great structures, knew how to

communicate with the atomic E-mail address

of the material they were working with.

It is clear the chief instigators of such
great structures were all cross-matter
accorded through advanced mind bridging.
With their lost art of matter communication
they could politely instruct the stones to
reduce weight. In truth, it still took
rollers, levers and muscle power, plus super
great positioning skills, but the material
itself was in a form of levitation. Their
art was to keep the smart side at a fine
constant.

Ultimate control of formal matter smart side
leads to the realm of true magic.

Back to jolly time travel. Lock onto the
planet's smart side and you will stay with
it during the time travelling. Of course, it
would be advisable to go into orbit before
materializing to avoid the problem of matter
convergence. A suitable disonic force field
around the vessel could be applied. This
would give suitable protection.

But any substance co-existing within the vessel's spatial axis would be pushed aside to allow for materialization. Nothing could resist the emergence of a shield protected time travelling device. Any substance, rock, sand, water, or a building would have to make way. This could prove rather noisy and messy. It would be as if a bomb had gone off. Not the most discrete way to arrive.

Waiting for the spatial ground level to be approximately level with the vessel before applying the breaks and observing that nothing co-exits within its space axis is also a rather sad option. Clearly then materialization is best performed in high orbit where the force shield only clears away the fine matter of space. It would still create a nice Star Trek whoosh of twinkle light.

But during actual time trans-positioning ground level observation is quite possible, and enjoyable.

Alas, our 60s time traveller didn't go into
orbit before materialization, staying at
'the same' ground level. Nor was there any
mention of a disonic force shield. For him
even a blade of grass would have proved
disastrous. The blade of grass matter would
have co-existed with the atomic structure of
the machine. Now you would have a big bang.
We must allow poetic licence. It is still a
good film and rightly a classic of its type.

Of course, it does not have to be formal
matter sent down the zint time line. Beyond
this, and part of ultimate advancement,
sufficient thought power can accomplish the
feat. To think is to go. However, once
achieved, by formal matter methods or
thought control, one has then opened a
channel, a form of thought particle beam,
along which formal matter can travel. Zints
are only one step above Universal thought.
It is therefore quite possible to send
thought, and its big brother mind, along the
zint beam.

This will be useful if we only wish to
observe events. Truly advanced civilizations
within the galaxy do this constantly.

The enlit have reached to explain. The
twelve truly advanced civilizations within
our galaxy watch events through a chosen
individuals eyes. In effect, they become
that person from the safety of their own
time and space frame. We have the beginnings
of such ultra advancement now. We send
robots down to ship wrecks. We send robots
to Mars, our origin planet. The military
have developed thought control for fighter
pilots. They become the plane. Soon, they
will become the plane but they will not be
in the plane. Military robots are being
developed that can be sent to fight under
the thought control of a soldier that would
only step onto the battle field to mop up.

We are at the horizon of such technology,
because we are meant to be at this horizon.

It is the way we are guided for us to dream
the next horizon.

The enlit have reached to explain that our
current attempts to communicate with extra-
terrestrials will succeed, but the given
communication will be very weak. There is a
universal mode of communication between
technically advanced civilizations. Digital
gravity. Pulsed zints. With their ability to
cross space and time instantly, the digital
zint wave will transmit everywhere at the
same moment. We must develop gravity pulse
receivers.

Mars is an extremely 'haunted' planet. There
are many trapped sentient life shadows,
still locked to the planet's essence field.
They have had no manifestation contact with
a physical existent life form for millennia.
The first to explorer the planet should be
made aware of this and be prepared for some
extreme forms of 'supernatural' event.

Time recordings and sentient life shadow
activity will erupt as soon as the suitable
playback and aerial mechanism of a corporeal
mind ventures to the planet. After the
initial shock the time recordings could
prove useful as a guide to the planet's
history. The sentient life shadows will
eventually settle down and many will be
eager move on to a higher state, after being
given the extra will to break the astral
chains by contact with physical existent
life forms.

Truly advanced civilisations watch events
through receptive individual minds. The
receptive would, as a rule, have no idea
that they a multi-quantum experience camera.
Sometimes however the thought probe can be a
smidgen too strong, or the receptive just a
little too intrinsically psycho-sensitive.
This is when awareness can occur.

It is a sudden feeling of 'All this suddenly looks very strange to me, but very interesting. I think I will have a look around'. Next time you see someone stood looking as if they've never seen this planet, wave at them. You could well be waving at minds thousands of light years away. You could well be a feature on their travel log.

It has been stated that true time travel materialization is best performed in high orbit. This is the basic rule. However, there is the possibility of extreme exceptions. To materialize at approximate ground level would require having precise knowledge of the planet's smart side. A reference point would also have to be known. It would have to be a reference point that was thought secure to remain constant for many thousands of the planet's years. Something built specifically for the job, which was sturdy enough to last millennia and provide a discrete arrival point.

Incorporated within the structure would be materialization chambers and means to align equipment against the planet's galactic star view, thus checking that the desired point in time had been arrived at. Narrow shafts engineered with great precision through the structures masonry. The chosen planet's rotational properties, its orbit and galactic procession would, of course, be known to the time travellers. These shafts would serve the duel purpose of allowing the air to escape, disrupted by the disonic force shield surrounding the emerging device. Air would also be displaced along the passageways within the dream beautiful structure. Such time stations do indeed exit. The twelve truly advanced thought and time faring civilizations spaced throughout the galaxy have placed them. They have been built with given wisdom, ultimately shining from the enlit.

It would be logical to build these time station structures on planets with known potential. Their design is always one of elegant simplicity yet time sturdy. They are built with the knowledge that their true significance will eventually become known. A process that is happening NOW. Their secrets are designed to be slowly revealed through the emerging technology of the planet's cultures. We know these structures as the pyramids.

To say that the pyramids are tombs is to say that Stone Henge is an ancient venue for a rock concert. The simple truth is that they cannot be built with our technology, yet they are thousands of years in existence. They are a calling card.

They could also be used for none formal matter time travel, as a focal point for the mind to arrive at. Once attuned the viewer could then send their mind out to observe.

Again, the shafts could be used as a
reference. If material travel is desired the
structures, as explained, are on hand as a
form of trans-time trans-dimensional bus
stop. Bus passes are available from the
enlit. They always were, and always will be.

WordLight Eight.

Light travels far for us to see,
Yet we do not. How can this be?
Perhaps the answer to this lies,
In opening up our secret eyes.

And then, what new sights to explore,
When once before we did ignore,
To rid the lingering shadows still,
That would close true eyes against our will.

AWAKENING THREE.

The enlit fully reached to me for a third
time just over a year later. During the in-
between time I could still sense their
presence with the occasional small star
grain of inspired thought.

They shone regarding the builders of the
pyramids, and the lit answer shone back that
WE did. Our distant descendants will become
part of the twelve and master the art of
thought and time travel. They will supply
proof by thought induced instruction,
resulting in the construction of structures
that can only be built using highly advanced
smart-side methods. The pyramids were never
truly intended to be entered on a pure
physical level. Although possible, the true
nature of their interior can only be
appreciated by access on an astral level.

Our technology is only just reaching into
what has been phrased nanno technology.

The pyramids represent the most supreme
advancement of such technology. The key has
been lost, but if it were found, the
structures would come 'alive' as intended.
What we see is a huge sleeping secret. The
very structure of it holds all answers. It
is as an ape man stumbling across an
unplugged television.

They also shone regarding the meaning of
life. The lit answer shone back. The meaning
of Life is LIFE.

The third full reaching was stronger than
the first two. I reasoned that they were
slowly turning up the volume. Again, I was
alone, walking through the park after a day
of happy playing with friends. I was in a
happy receptive mood. When I fully sensed
their reaching, I 'instinctively' found a
quiet grassy place surrounded by lush
privets, rhododendron, and hydrangea bushes.

It was as if the private golden sunlit space had always been waiting. It then occurred to me that the instinctive finding was nothing of the kind. The enlit had guided me. The enlit had probably created the thought space, instructing the thoughts of the park gardeners to leave a secretive grassy area surrounded by flowering bushes. And by thinking these thoughts, the enlit were telling me the truth. I sat down and calmed within the grassy circular sunlit flower walled thought chapel.

The spark flew. Their reaching went for the first time into the nature life of gravity. The gravity particles hold the structure of matter together, charging negative and positive accordingly and creating the space we perceive by the nature of their motion, which in turn inspires all temporal material particles into space creating motion. They also create the time for such spatial motion.

For the Universe to be aware of itself on
all levels it must be in touch with itself
at all levels, and it must have memory of
itself at all levels.

On an infinitely smaller scale 'you' have
the memory that your hand exists, and it
still exists should that hand be physically
lost. Your mind contains the memory. But the
memory is in fact a residual trace gravity
recording of a hand that gravity has held
together. Your quantum mind filters out all
overloading colours to leave the basic
impression of your hand. It filters out the
finer details of what your hand's inner
structure is or was doing, and also what the
whole did in the past and will do in the
future. Your mind only retains a memory of a
'now' hand. There is no point in overloading
you with a myriad recollections of the hand
structure movements through time and space.
A 'now' hand is quite sufficient for our
everyday mind to cope with.

But the gravity particles also have a trace to your future hand as well as your past hand, being able to hold matter together across time as well as space. Should your hand be tragically lost your mind will expand slightly and the filter will allow a partial amount of the gravity particles experience of your past hand. You would start to experience memories and recollections of your hand.

However, the gravity particles will have travelled to a future in which your hand does not exist due to the future tragic accident. The fundamental wisdom of the Universe, its servant Nature and via Nature the enlit is to expand the light of awareness slowly. Sometimes the wisdom is itself expanded and awareness is given an extra spark. The filter created by your mind expands beyond the precedent horizons. It allows a zint gravity particle awareness of a future 'none' hand.

This is manifested to your awakened will consciousness as a premonition, or also precognition.

Premonition and precognition are a mind expansion of the mind's zint gravity filter to allow more of the zint particle time experience to register on our level of material physical consciousness. The zint particle carries the 'memory'. This memory is an essence, a particle of consciousness, an awareness of that time moment, and a none physical statement of self existence at a different universal point to the one 'now' self experiencing. It is a statement of being. Being that exists past, now and future, which is how our level of mind divides time, for its own convenience, and to prevent it from loosing itself. It could then become too matter-and-dimensional.

Zint particles transmit thought awareness of the whole across space and time.

The matter sentient Universe knows what it
has done, what it is doing and what it must
do. But at the Universe conscious level,
these are fully trans-happening. Zint
created gravity allows the Universe to know
that it IS what it IS. Zints are the mind
ball-bearings of the Universe at work.

The fundamental wisdom of the Universe is to
allow awakening mind the occasional glimpse
into those workings. It allows our evolving
mind filter access to zint gravity particle
cross time experience. It allows the enlit
spark of premonition and precognition. In
most it is a fleeting experience. In a rare
light ready few it is endowed as a true gift
and the mind filter is on semi-permanent
acceptance of zint created gravity particle
trans-time experience. These gifted have
premonition and precognition. They are the
psychics. And yet, the fundamental essence
wisdom is not perfect and never can be.
Perfection is unattainable.

To continue polishing the mirror will eventually polish through the mirror and destroy it. An element of imperfection is an essential to perfection. True perfection will have the last element of imperfection. Take that away and perfection becomes meaningless as there is no reference point for perfection to relate to. Everything positive must have a negative element to relate to. When the time falls and there is only one true negative spark element remaining, perfection has been reached.

The Universe still has much thought to travel to gain experience before pure perfection / none perfection is reached. Unfortunately the enlit wisdom of allowing mind filters to accept zint gravity particle trans-time image can shift in the benign prism and allow the gift to fall to darkened physical and mental templates. The gift is then used for the spirit negative aspect which is darkened as possessively controlling free will.

WordLight Nine.

What treasures lie waiting to be found,

When mortal minds awake,

Wonders still invisible lie around,

Willing us to take.

All knowledge lies a dream away,

Ancient shadows fall between,

As we in sleep pass through each day,

And miss what we have seen.

The true light ready, the true psychics, use
their gift of unfiltered zint gravity
particle trans-time experience discreetly.
Alas, their call is often too quiet. But
this is Nature's way. Would we love a rose
if it were six feet across? It is the fate
of the light ready to dull their shine. But
shine they still do.

Premonition, precognition, affinity with and
memories of the perceived past, beyond the
physical lifespan, are tiny fractions of
quantum mind filter opening to allow poly
dimension zint gravity creating particle
trans-time experience. Past perception is
also astral coloured by the zint particles
remembering their matter control of a past
life 'you'. And it is the 'you' in all
things that makes existence possible.

The reaching closed.

I returned out from the grassy circular
sunlit flower walled thought chapel into a
darker but still beautiful sunshine. It
wasn't enlit. But I was now beginning to
truly sense my own enlitness. I knew I was
possible. The glow was becoming perceptible.
It had been a good reaching.

The thing called Nothing is totally and
Universally smart side accepted as pure
impossible.

llumination Seven.

There is no such every matter conceptual
thing as nothing. Nothing is a concept
created by creative thought to perceive what
is perceived as static quantum-time-space
activity. Mind has to have something to
relate to, so it creates the concept of
nothing to allow itself a breathing space
between events. If something is created then
it exits. One cannot create nothing.

The Universe did not create from nothing.

If thought can conceive something then all
that remains is to discover the advanced
physics and mechanical means to achieve the
physical realization.

Somewhere else is here.

All things are with us throughout the mind continuous moment. If the thing we are looking for is somewhere else, then so are we. We are the people who are looking also. It is the thought-time filtered through sentient consciousness which gives us the spectral multi-variance of experience. The things-being-apartness we perceive as reality is in fact a reflection of the whole. Thought-time creates the fractionism we need to exist and allows the laws of reverse positioning. Reverse positioning law prevents zints from becoming too time and space active. If they had total free will then they could expose us to our past and future at random. Nature controls zints with the laws of reverse positioning.

Nothing can travel faster than the speed of thought. Thought is everywhere and at all times through every level of space. One cannot travel faster than being everywhere at the same time.

To go faster is beyond thought to conceive thought, and thought can always conceive thought. To go faster than thought is a none impossible. Thought is eternal. Light is the shadow of thought.

The enlit have shone regarding the essence of experience. Experience is the dream behind the colours of a rainbow. It is the will of light. Can light see itself? Only by a conscious entity. Light is something by which consciousness can see. Light can have no idea what it looks like on its own. It needs 'you' to show it. Our collective thought allows us to do this. The greater thought produced the light for us to experience in the first place. Everything wants to experience itself - through multi level thought.

All movement is outward.

Direction is always towards the point of origin.

A point of origin 'is' one. Two objects may seem to pass each other but they are in fact returning to a common point of origin, as all C particle matter is. Place some diced carrots, parsnip, potato whatever onto a plate. Add a small amount of salted water. Cover the plate with cling film. Prick the cling film in three places with a fork. Imagine we are an atom within the cling film. This is stretched to have contact with the edge of the plate. The edge is a continuous curve, and no matter where one starts from, the direction is outward, i.e. away back to yourself. The atom, 'we', have remote contact with a continuous away-back curve. The plate is placed within a microwave and the cling film will rise like the millennium dome. We atoms note that space is stretching in all directions, yet keeping contact with the continuous curving 'edge' of the plate Universe.

This is a one dimensional analogy of course.

The true Universe has an infinite number of
'edges' aligning through an infinite number
of directions. This would require an
infinite amount of cling film. An infinite
amount of space i.e. edges on which to hang
an infinite amount of away-back starting
points for thought to travel outward from.
The Universe spuds take about eight minutes
on full power to cook thoroughly and become
well boiled. Allow five minutes to parboil
the galactic potatoes and they will be ready
for deep frying.

The Universe is boiled potatoes.

The beam of a lighthouse sweeps across the
sea. The lamp revolves every ten seconds.
Close to the light source the light beam
sweeps through a matter of yards within the
space of the ten seconds. Out to sea the
light beam has moved many miles, within the
same time frame.

If the light beam was solid and you were sat
in a capsule fixed upon the end, your speed
would be many thousands of miles an hour.
This is a quantum relativity paradox. The
speed of someone in a capsule fixed close to
the light source would be infinitely less
dizzying. It would be just a few yards in
ten seconds.

A gramophone record is solid yet, as it
turns, its solid material edge is moving a
greater distance then its allegedly solid
material form close to the centre, within
the same time frame. I sit in a swivel chair
on top of a hill. There is a panoramic view.
I turn in the chair. My body moves in a
circle of only a few feet, but my vision, my
line of sight, moves many hundreds of miles.
My thoughts have travelled a great distance.

Imagine if it were possible to hold a rod
out from the conscious planet and into
space. The rod is many light years long.

We must assume the engineering and material problems have been surpassed. Take away atmospheric resistance. Take away the fact that that planet is revolving at aprox. one thousand miles an hour and that it travels in its orbit around a star which travels within a galaxy. In fact, imagine that you and the planet are stationary.

Forget for the moment that this is a non impossible, as thought is movement in consequence. Now, you point the rod at a star. You move the rod to another star which appears close by in your sky. The end of the rod has thus travelled many thousands of light years in just a few seconds. These examples maybe somewhat simplistic in their objective but they serve to give a glimpse at the quantifluxing nature of time, space and material C matter. The fact that a record can turn, that its substance can move through greater distance along a fixed time frame, is proof of dimensional flux.

As with the light beam and the rod, their core essence floats on a raft of non time perfectionated resonance. This itself is held in harmony on the super purveyance of thought.

Movement is a non voluntary response to co-existanal spacial interference.

All things must move. Spatial interference insists on this. Nature is movement. Movement is an essence of Nature.

Still does not exist. To stop is a non proportional perspective against the proto-linier time frame of natural thought. Things move to evolve and all movement is the evolution of one form to another, one idea to another, within the multi-directional ideas of essential thought. The Great Thought wants a dolphin. The idea of a dolphin then moves. A dolphin is created. Within our time frame this takes many millions of years. Within the essential time frame it is instantaneous.

To back lineate on this, movement is
existence. On a much smaller scale movement
has to be to provide the correct balance
within co-existent spatial interference.
Conscious level one material matter does not
like to touch. Only when it combines to form
a sentient higher consciousness does touch
occur without explosion. As a rule, the
higher the consciousness, the greater the
degree of move augmented touch occurs within
its space and time frame. The movement
becomes as of one and not as many.

Distance is a fact of thought. When a
distant galaxy is seen for the first time,
we have in fact reached it. Thought has
perceived it so we are immediately there in
thought. The apparent action of travelling,
to a galaxy, Blackpool, or the corner shop,
is simply the creation of more thought to
take us closer to where we already are. We
perceive and we are there. We have thought
to travel. The apparent action of travelling
increases the amount of our being there.

As always there is positive and negative
being there. To move 'closer' to where we
already are is being there in its positive
form. To move 'away' from where we already
are is as being there in the negative
aspect. It is this consequence which gives
rise to the enlit hypothesis of revolving
fluid reality. Their reaching illuminates
through zints that I should not be here.

A pebble is thrown into a still pond. The
ripples spread towards us and away from us
at the same time. If one moves through space
and time to the other side of the pond,
those ripples that were moving away are now
approaching. Yet we have moved around the
bank of the pond - which has continuous
connection with the pond, and the time and
space of the pond. Now imagine you are on
the ripple. Are you moving towards or away?
It takes a second diverseotonic viewpoint to
see true events.

It is the viewpoint needed to see the person

who threw the pebble creating the ripple.

You can either be approaching them or moving

away from them depending on which side of

the pond zint given time and space allows

them to be. On a Universe infinite pond the

ripples would eventually spread and fold

back on themselves to return to the original

impact point of the pebble. Light will

eventually return to its point of origin. If

reflected in any way, a mirror for instance,

it will return to the mirror, which itself

is returning to its own point of origin.

Should the zint charge light photon hit

further mirrors, or points of reflection, it

will have a new return point and so on. But

each return point is returning to a

universal original return point. No matter

how many times the light becomes reflected

it will eventually return to the one return

point of all things.

We look into a mirror and see ourselves

reflected.

Light is radiated from us and returned. When
we look out into space we are actually
looking from within the mirror. We are the
reflection. The greater the distance out,
the greater the distance within the
reflection in the mirror. The further one
looks out into space the deeper one travels
into the mirror.

The distance in-between our conscious being
view point is the space and time thought
that the Universe has created for itself to
Universe happen. Thought is the water on
which the perception ripple moves. Thought
is the medium of all reflection.

There is no such thing as destruction.

Everything is in a state of creation. There
are constant states of positive and negative
creation. This relates to positive and
negative thought.

Thought can never be destroyed. If it could be destroyed then you would not be here to give thought to reading this, which is a creation of thought. As all things are ultimately thought, we are potentially all things. Thought cannot create the thought to destroy itself. This would be to create nothing and, as we know, jolly nothing can never be jolly created. Creation is all happening. Creation is thought, and zint created gravity is the sentient creative shadow of thought. Light is the shadow of gravity, by which the Universe allows us, and via us itself, to visually experience all physical matter. Zint created gravity curves space and tells us the mass of physical C matter. It tells us how deep we are. An echo sounder tells us the depth of a body of water. We send the echo. Our thought needs to know the depth of water. Universal thought need to know how much matter it thinks. The echo of its questioning we perceive as gravity.

Nature dreams all things with two aspects.
We perceive this as positive and negative.
It is the same with gravity. It can be
positive and in can be negative, due to the
will of the zint particle. However, the zint
can be persuaded to create negative gravity.

To persuade zints to create negative gravity
(anti-gravity) for the purpose of advanced
propulsion systems is a weighty subject that
requires careful thought. An early method
would require first and foremost the
creation of a pure diamond alloy. To
simplify matters a base away from Earth is
needed. Once created this substance would be
thousands of times harder than its naturally
forming counterpart. A sphere of the diamond
alloy must then be suspended at absolute
zero temperature within a poly-plasmic
containment field generator. The housing for
this 'must' be lined with gold. Copper, tin
or lead will not be adequate.

Gold is a natural shield against space polarity distortions, or, negative gravity exhaust.

We can now see one of the reasons why the Ancients worshipped in gold. The sphere is set to spin. Any form of spinning creates a warping of zint particles with an inducement to create negative gravity. But the spin needed for this form of propulsion is very great indeed. Again, the substance spinning must be absolutely pure and it must spin at the exact resonance of its sub atomic nuclei, possibly many trillions of times a second. This is the secret. When the resonance is reached the zint particles themselves are 'thrown off' by the terrific centrifugal forces. They are 'fooled' into thinking that they shouldn't belong here any longer. What is actually happening is simply that the curvature of space caused by zint movement is curving back on itself. It is as a ball turning itself inside out. Gravity then becomes a mirror image flying outward.

Negative gravity has thus been formed and the alloy would instantly loose what we perceive as weight. This would then take the generator, and the ship, with it and away from itself.

Of course, the negative gravity has to be focused, or geared down, to control movement. This is done using omni fluxing quark prisms. These are quarks that have had their polarity removed so that the zint particles can reflect from them. The tiny spatial distortions, 'ripples' from the exhaust which can be created, give rise to the halos of beautiful colours and dancing lights seen around some U.F.O.s, especially those which are only a few decades ahead in technology. The truly advanced misnamed U.F.O.s (quasi-matter intransitals) use time cohesion plasmonics and their colours are actually seen (daydreamed) only within our minds. Negative gravity created by sub atomic resonance can also be stored, just as energy can be stored within a battery.

The battery in this case being a form of space envelope (quarto) created by a super conductive ionization field or a baby plasmic containment field generator. The stored negative gravity can then be released under control. They could also be directed through quark prisms within the structure of the ship to create the classical idea of artificial gravity.

An alternative solution to anti-gravity, would involve the creation of a hyper gravitational duct, a cheap form of black hole, ahead of the vehicle. Imagine a man on a trampoline with a garden roller. He pushes the roller across the trampoline creating a bend in the fabric on which he stands. Man and roller fall into the dip. The vehicle creates a dip in the trampoline of space. It pushes a 'garden roller' of artificial dull matter. A gravity 'puddle' is formed and the vehicle is constantly 'falling' into it. The vehicle moves from puddle to puddle across space.

Although dull matter is relatively easy to produce, by colliding the sub nuclei of helium within a multi fluxing magnetic field, the real difficulty with this method is controlling the size of the puddle. Sooner or later the man would have to let go of his roller.

A less messy and more sophisticated form of anti-gravity propulsion is the elastic band method. The destination point is targeted. The space in between is 'stretched' along a suitable zint carrier wave. The 'elastic band' of space is then released and the ship is catapulted along a controlled trajectory. It is a form of gravity catapult. This is complimentary to the better still gravity squeeze method, where the actual space 'travelling' is reduced to a minimum.

Two points along the time spatial and gravitational matrix are bent so that they almost touch. Imagine the surface of a bubble.

A small area on each side is drawn inward, like glass being blown, so that they virtually meet at the centre. If they did meet, our hypothetical space bubble would rupture. In reality the two points of space would lock. Space would flood through from one side and then be washed back again by a flood from the other side, like an inter-dimensional wave machine. The amount of space on either side of the hook would always fight to remain even. To be trapped within this anomaly would be like being trapped in a revolving door, at great speed, whilst wearing a diving suit. Such would be the discomfort. Space is always perfectly balanced throughout the Universe. One part could never be flooded by another part and start to sink. Space will always rearrange itself to stay balanced. However, the two 'bubble sides' of space which we have pulled together, i.e. the two points one wishes to travel between, are kept apart by a negative space vortex generator.

This is, of course, the 'fictional' portal
that we all know and love so well. A
Stargate.

Beyond all this is the above mentioned time
cohesion plasmonics. Thought generates time
so that it can have space to think in. If
the vehicle can think hard enough about
going somewhere and sometime, you go. The
vehicle can either be constructed, a living
entity, or, oneself.

As you can imagine, this is a very advanced
form of transportation and realization lies
several generations from this book point of
origin. But it will exist for us and does
exist for others, because it can be
imagined. The concept, as with all concepts,
is a product of thought and if the focus is
strong enough the concept will become
material form. The thought will become real
and join all the other reality thoughts.

The enlit have reached and shone to thought illuminate the varieties of U.F.O. concept.

Planet static. Random discharges of plasma.

The advanced, faster than light, space warp vehicle, with occupants. These are actually only a hundred or so years ahead of us in technology. We are catching up fast.

The fully trans-time trans-dimensional vehicle, with occupants. These are a little further along our technology motorway, but not too far beyond the horizon.

The 'SPACE FISH'. These are entities in their own right. They live in space as fish live in the sea. Their shapes can vary, as does the shape of fish. Some could be regarded as 'space birds'.

These entities feed on gravity glow, the disturbance of space caused by zint movement creating sub light particles. Space plankton. These entities have been witnessed creating swimming and shoaling motions.

The thought probes. These are the projected minds from the advanced civilisations existing within the galaxy and beyond. They are their mind robots, just as we would send a robot to explore a wreck on the seabed. They send their mind robots here.

The astral entities. These are slightly more sophisticated as they send themselves, but in astral form. The form that we experience will have no resemblance to their true form. We would experience the classic u.f.o. multicoloured glowing ball. Occasionally they do choose to send their astral form in a recognisable human shape. Experiences of these have been inspiring.

Gravity lightening storms. Zints caught in a
zint traffic jam.

Planetary consciousness thought fusion. We
witness the planet dreaming.

Space ghosts.

Multi-dimensional Universe thought sparks.
We witness the Universe dreaming.

Thought lights from the enlit.

WordLight Ten.

As Nature bids us fond farewell,

So many stories yet to tell,

And if we lost all word of it,

Still, in Nature's dreams, our story would

be writ.

WordLight Nineteen.

To touch the sky is but a dream,

And in this state above we seem,

Midst all frantic ghosts of memories gone,

Still we stay to carry on.

llumination Eight.

Somewhere and sometime are as the surface of
the pond. Two realities, above and below,
separated by some mere reflection. Now ask
yourself, just how thick is a reflection?
Matter is space trying to get out.

We try to discover the smallest particle of
matter by smashing it together. All that
happens is that we break the dinner plate
into ever more smaller pieces. The zint
cannot be found in this way. It will not
allow itself be found in this way. It
doesn't need to be found in this way. Its
existence is proved by the gravity it
generates through its trans-dimensional
motion.

The pieces of the plate will become smaller and behave in stranger and stranger ways only leading to more and more questions. Eventually, the dinner plate will smash with perfection. We will physically experience particles which defy all physical time and space as we materially know it. It is these any-timers which are almost the final fragments of material form. But they are not zints. It is the zints that give them the qualities we will marvel at. Zints are not part of us. They belong to the whole. The zints are so none particle in structure they are time fluid. If we ever did In discover a way to view them we would find ourselves at the very limits of physical monitoring. Zints are the line of the horizon. They are the surface of experience. Beyond, the ultimate building substance can never be seen because thought cannot be seen discovered or experienced with raw matter. But substance is a misleading word. It implies particles.

Thought is none material matter. It is the space within the soap bubble. Zints are the surface of the bubble.

The past and the future are always one. Time is a constant N.O.W. A Neo-experience Over-thought Webb. We exist in concept at various points within it. What we perceive as formal matter, our Natural Universe, exists on a fluxating platform of zint particle essences accelerating though time points along the spatial and inter-dimensional interfaces we call reality. It is the zints which hold a key to Time Travel in the classical sense. However, their control will be extremely difficult, but not impossible. Nothing is. But, with zints, nearly so. The difficulty lies in the fact that they can never be in one time-space spot for more than a trillionth of a second. This is not in their nature. They are meant to flit from time point to time point, which, from our point of view, could be perhaps millions of years apart, or merely seconds.

To them it is meaningless. To elucidate
further, think of a television screen. The
set is switched on but as yet isn't tuned in
to any particular channel. We see static on
a television screen. This is formed by
countless electron particles. They exist on
the screen for a tiny fraction of thought
time but there are so many of them appearing
then disappearing that to our outside mind
eyes a continuous swirling mass is created.
This is an approximation of the reality that
we experience around us. The electron
triggers a pulse of light which we see. The
zint creates a point of existence from which
a material particle can be charged to
manifest. Counterfactually, the afore
mentioned electrons creating the static
light pulses are themselves existing on the
'static' of zints on the slightly bigger
multi-dimensional poly-time screen of the
Universe. But Nature has provided a way.
Nature always has an answer waiting to be
found.

It is the nature of zints to flit through time and space. It is their job. However, in doing so they loose a tiny fraction of their own energy / mass. This is recharged upon the moment of materialization, only to be lost again on the next journey. A continuous process of mass loss and gain as they move and change from negative to positive as required. This accounts for some of the so called dark matter currently being raved about. However, the recharge is never quite the same and it is this un-sameness, the amount of energy forested, which can influence the distance the zint travels to the next point of manifest. If it becomes aware of a particle in need of a negative charge, but it is a million years away and the zint feels it may not make it, it will choose a closer particle, knowing that another zint will rescue the first particle. If it were not so, if the recharge were always at a constant, there would only ever be two points of space-time, with no change and nothing in-between.

It would be a frozen NOW then and a frozen
NOW now. Now, to have a frozen NOW now and a
frozen NOW as will be, or as was a now
before it became a now then, can now be seen
clearly, although now knowing indirectly, as
a now absurdity. NOW always moves.

The static on the screen is in constant
motion, as is the static motion of our zint
floating reality. The zints must loose and
gain whilst flitting. It is also one way in
which the Universe distributes it's mass of
reality evenly throughout space-time. It is
here where a key lies.

The amount of energy forested determines the
distance the zint travels through the space-
time continuum. What is needed is a zint
gathering device and a process for measuring
their energy mass. The heart of such a
device will itself consist of zints,
amplified and tuned to seek out other zints
with the chosen properties. It is in the
nature of things that zints can find zints.

Like can always find like. Light can always find light, because it has the brightness to see where it is going. Darkness will always find darkness, but it takes a little longer to find your way in the dark. This is how Nature has applied the balance.

Now it will fall that a significant proportion of zints will have amassed the same mass and will thus manifest together at the same time reference. Choose zints with a collective low energy mass. Herd then all together and tag them with a small amount of mesotonic radiation, which is extremely rare but easily identifiable. It is given out by dieing mesons as they approach the event horizons of black holes. The tagged zints are then released to do their own thing. Here is the pain. We scan for traces of mesotonic radiation that have appeared in the past. If nothing is found, we wait and constantly scan for mesotonic radiation to appear in our future.

Either way a bench mark is found regarding
energy equated to distance travelled through
space-time. The art will lie in choosing
zints with a low enough energy so that the
wait, if it has to be, will not be too long.
Of course, if mesotonic radiation is found
in something dated in the very recent or
very distant past a guide will be
established as to how long to wait for it to
manifest in our continuing future/present.
The next trick will be to 'point' zints into
the 'past' or the 'future'. But these are as
one. To progress it must prism from the
enlit an explanation of the principals of
now-ness.

As stated, time is a constant NOW along
which our consciousness travels through a
reality dictated by the constant holographic
manifestation of zint particle essences.
They are so small as to be almost the
thought essence of substance. To 'point'
them past or future is to establish a
unified flow of space-time consciousness.

This would be a thought zone that has yet to
be or has already been experienced. We
experience a permanent now, but there are an
infinite number of nows which could be
reached. The so called past and the future
are happening now and always have done, but
the nows are not happening in the now that
we currently occupy.

You drive along the famous motorway at speed
and see a bridge to pass under in the far
distance. The road ahead of you exists in
your mind and leads to the bridge, in your
mind. The zints have created the gravity
knowledge for you to know. You 'know' you
are travelling along a motorway. Now, close
your eyes. You still 'know' that you are
travelling along the famous motorway, but
you can no longer see it, nor the bridge
ahead. You can still experience the motion,
but you cannot see its greater properties.
The zints are now allowing you just an echo
of their qualities.

You can imagine those properties, from all
the gained motion experience, but you cannot
actually see the road ahead or envisage
where it leads to.

However, sometimes the wisdom of Nature can
allow the eyes to blink open and we glimpse
the road ahead, that we had zint thought
imagined. The future nows become visible. In
some the blinks are frequent. The psychics.
Zint tuned satellite dishes for the enlit,
although they may not be aware of it.

The past and the future are as reflections
in two mirrors facing each other. We are
forever at a central point of cross
reflection. The mirrors change focus, i.e.
our consciousness appears to travel through
time, but we always stay sandwiched between
the mirror surfaces of perceived past and
present. It is the privilege of the zints to
create the formal matter framework so that
we will always have a now to experience
through our thought process.

The zints create the surface of the mirror. Our aim with Time Travel is to place a small prism in front of the refection. It is to focus that reflection and capture another now.

Once a now has been experienced it becomes negative. A now that has yet to be conceptually experienced is, as one would expect, positive. We remember negative nows in the form of memory. This is how the quantum brain has been programmed throughout its evolution. It is how Nature usually wants it. It provides a constant point of reference and avoids confusion. However, sometimes the mind is allowed to remember positive zint charged nows as yet pre-time waiting to be experienced. This is known to us as premonition. We have experienced a little reflected light from the prism. One might think that this pre-experience of a positive now would eludeify it into a negative. It does not. The reality of matter is that the now becomes a neutralite.

Only a full conceptual experience would passidicate it with negativity. It is these pure transient pre-experiences of thought which create sufficient neutral nows to provide yet another essential balance to the space-time framework of the ever thought Universe.

Returning to the future history of pointing zints. We wish to stream them towards a positive or a negative now. To do this they must be empowered with a positive or negative multi-directional spin. Remember, zints are particle essences. They are the hooks on which formal matter hangs. Their physics are totally different. They can move three ways in sync. A positive or negative spin will simply get them to quantify their time-spatial position. Imagine a field of gyroscopes. A huge magnet then flies silently over these independently spinning toys. Suddenly their oscillations become unified and their spin points a certain way.

If zint created gravity free they would flock in that direction. They would flow towards a 'past' or a 'future', as do our matter-holding-together friends the zints.

Birds, fish, insects, form a mind link to flock, shoal or swarm. They temporally become as one entity. Higher animals form packs. Humans go to football matches.

Pointing the zints? Well, correctly focused anti-nelons from dieing antimatter has been known to achieve this. Containment fields will have to be incredibly powerful, or a suitable multi space warp incredibly dense, to withstand the space-time pull. We must then make sure that the smart-side phasing is suitably shielded from distortion. Zints are the radio wave and formal matter is the message. The smart-side is the base and treble of the music. It would need a proportion of zints to securely gravity wrap the smart-side programming and protect it from the flux of dimensional warping.

Time travel is thus accomplished. Suitable applause from reader audience to the reaching enlit. We now have a Time Machine in the classical sense.

What it looks like depends on how the style team can wrap up the mechanics. The art of nano technology will help considerably, as will multi-phase and quasi-plastic smart materials, i.e. materials that can pretend look like other materials.

For instance, suitably shaped and decorated rods of brass which are actually trans-dimensional stabilizing coils. Leather seating which proves to be ditrainium power cells. With such materials it is quite possible to build a Tardis. The internal dimensional expansion is an option, although this would prove difficult given the Tardis shape profile, and dimensional warp generators are frighteningly expensive to make from quasi-plastic material.

Another whimsy would be to construct a time
machine which looks exactly like the device
in the classic 60s film. This is quite
possible with multi-phase quasi-plastic
smart materials. If the mind can imagine it,
then it is possible. Because 'you' reading
this, as with many others, can now imagine
such a possibility now increases that
possibility. Imagination is how the Universe
dreams to improve itself.

A 60s time machine would be somewhat
draughty perhaps. It would need the
refinement of a suitable enclosure field,
environ controls, and a seat belt. Once
taxed and insured, and a suitable turn of
the century costume acquired from a good
theatrical supplier, you would be on your
way for some jolly time travelling. However,
the chosen shape for time machines is the
conventional tried and trusted flying saucer
arrangement. It offers the better cabin
space for ones camping and picnic bits and
pieces.

It is less windy and has windows to wave from at startled Romans. They also look good in photos. Some chrononauts do go somewhat over the top with their craft, fitting a myriad of lighting effects. They don't actually mean a thing, except to have you, their 'in the past' observers, wide eyed with wonder and running around for the camcorder. Some time travellers have been known to scan the archives in search of photographs, recognised their own craft and then gone back in time in order for the photograph to be taken. One relatively new trans-timer had the misfortune to get himself photographed. He was quietly observing a man taking his daughter for a rural walk. He stood behind the girl as she sat on the grass ready for the happy family snapshot. He did not realise that the power reserve on his vision-mask was fading and that he would register on the film emulsion. To add to the predicament, he was from our distant past and was therefore in his own future.

It was only when he continued forward and his event then became recorded in his history did he learn of his error. An unauthorised time line had been created. He was banned from trans-timing for a year and had three points on his licence.

Time is the sea in which the universal conscious dream swims. It is the wave for mind to surf.

Our human species is not indigenous to this planet. The handful of survivors from the Mars planetary collision left their genetic marker many millions of years ago. Through us they survive. The pyramid builders and their counterparts who constructed great and mysterious engineering feats, Stone Henge, razor blade precision positioned massive blocks of stone at mountain top heights, all had thought links to the lost civilization. The genetic knowledge of controlled smart-side being a small jewel from the advanced technology genetically seeded.

WordLight Sixteen.

To dream is to create anew,

Realities that may come true,

For without dreams we stand to fall,

They are the wind that sails us all.

Teleportation =

Temporal Relative Interactive Particle

Expansion.

Space always knows where it is. It knows

because of the zint 'memory' which carries a

tiny fraction a space point awareness from

one point to another across time and space.

Infinity is only the distance between the

thought starting points. It is the amount of

thought awareness contained within the zint

that creates the time it takes to travel

between those thought points, which then

creates the space for the travelling.

Infinity can be a step away, or a trillion

trillion light years. The greater the

thought, the closer the apparent distance

point. The weaker the thought and the

apparent distance increases until we have a

perception of infinity. If we learn to focus

our thoughts inwards onto the zints that

hold us together and lock onto their event

memories, our thoughts will expand with them

and we shall thought swim to infinity. We

shall then discover that it is next to us,

and that the dream in-between will be

beautiful.

Zint created gravity is the astral milk

which feeds all temporal experience. It is

the food of the Gods, given the more poetic

name of Ambrosia.

WordLight Fifteen.

The mirror you, is it thinking too,

Does it ask, who is that out there.

Are thoughts reflected, trans-injected,

From the mirror into your somewhere.

For if so, it should know,

That you are also the reflection of it,

And for the reflection to be, there must be

life in me,

So which reflection has more life and wit?

Time is a pure example of none existent

reality. The reality of time only exists

because other realities allow it to exist,

chiefly, the reality of the zint, which

itself is the veil between matter and

thought.

We have thought and by doing so create time
to have the thought. We move and by doing so
create the time to move. Time could not
exist without a reality for it to exist in.
Time's none existence is proof of our own
existence. Our own existence is proof of
zint existence.

To exist is to create, and we create the
time for our existence. The artist creates
the sculpture. The sculpture then exists to
create time for its own existence. But the
greater thought behind the sculpture created
the time for the sculpture. The sculpture
then contains our memories of its creation.
The secret of time travel is become as one
with those memories.

Nothing is forgotten. Every particle of the
Universe contains a fraction of the memory
that created it. You are the proof. Your
memories are a spark reflection of the
Universe memory. You are thought time as the
living Universe is thought time.

We have learned to ride a bike. We will
learn to ride time.

WordLight Seventeen.

Fear never, time is for ever,
Those of woe claiming not so are beyond
reason's bright brink,
For if Time, so exalted, could actually be
halted,
You would still need sublime Time to woe
think.

llumination Nine.

Experience =

Matter Arranged Dimensionally.

The Supernatural isn't.

Ghosts etc. have always been a puzzle, being
somewhat aloof. But once the correct amount
of thought has been applied it is relatively
easy for their truth to be illuminated. What
is termed the supernatural, isn't. Nothing
within the dreams of Nature has any radiant
supernatural quality. All things are a
manifestation of Nature, and Nature is the
most natural thing there is. There is
nothing super about Nature, for it is
itself. The 'super' element is merely a
natural manifestation that has yet to be
fully illuminated by rational conscious
mind.

150

The enlit are now allowing some

illumination, which ghosts will not

appreciate as they guard their qualities

zealously, that is the sentient ones do.

WordLight Eleven.

We dream our dream then pass on by,

All our hopes, thrown to fill the sky,

With colours, of which we can scarcely

guess,

And if asked to fly to them, our answer

would be, yes.

For we cannot stand in wonderment too long,

Nor forget the wording to our song,

Lest all our gains should turn to dust,

We are our dream, so dream we must.

Sentient life shadows will be a little annoyed at having their secrets revealed. They will have to complain to the enlit. It is they who have decided that a little thought-light should shine on them, again. Thousands of centuries have past since the last great thought-lit age, when matter communication had been perfected and true magic created a thought garden world. The memory has filtered down to become the story of Eden.

The sentient life shadows know that the circle is completing again and a new thought-lit age is beginning. Their ages of secrecy now fades and they realise they will soon be fully understood, as lightening is now understood. They are loosing their supernaturality, which they never had anyway. As for the time recordings and dizzy roaming after glows, they are as witless as they have ever been. As witless as any form of none entity recordings.

Would you talk to a c.d. or to the shadow of
someone? No, you would talk to the person
who made the recording manifested on the
c.d. and to the person manifestly casting
the shadow of their life existence. However,
if the entity will was strong enough to cast
an element of that thought will into its
shadow, a shadow communication enters the
conscious viability. But there is nothing
supernatural about such viability. Is there
anything 'super' about you now having the
thoughts of such viability? All thought is
natural to Nature, be it negative or
positive.

The sentient life shadows will be rather
upset that the super tag to their phenomena
is now being challenged. Thoughts on them
are slowly being illuminated, partly through
this enlit inspired digest. The time will
fall when their manifestations are regarded
a mere nuisance. 'Oh no, we have mice.' 'Oh
no, we have ghosts. Get the spray out and
zap the buggers.'

All phenomena are natural. They must be so
in order to exist in the natural thought
inspired Universe, which has been around for
quite some thought time, so it should know
what it is dreaming by now.

'Oh no, we have ghosts.' Well, it is easier
than calling out 'Oh no, we have holographic
cross dimensional interference and poly-
axial de-fragmented reflections of a space-
thought-time event.

Ghosts have a variety of manifestations but
they are usually: Dimensionally Alternating
Fractal Telepathy, with manifest actions
that illuminate them as being: Poly Relative
Astral Thought Transmutation Signatures and
Dimensionally Interactive Cognisant Knowing
Holographic Eventual Actuality Diverse
Spectres.

My apologies to most sentient life shadows,
but you have only yourselves to blame.

The sensible controlled thought ones do exist, as would someone sitting in a Porsche, without having a single thought wish to drive it. They are happy playing with the electric windows.

On the perceived supernatural aspect of Nature's dream preoccupations, ghosts use very little energy. They are as a microwave, working, but with nothing to cook. Creative, Universe enriching thought has gone, until the microwave wills itself to plug into the next socket, higher up the wall. They are none light light bulbs.

These are the higher form of ghost. They still have the spark essence to interact with the reality they have left, as yet unable to absorb the new reality that they have entered. They quickly learn how to manifest their presence, realising that they can take hold of a material objects 'extra side'.

This is the level of every material form
that exists at a higher plane to our utility
material senses.

We have the visual joy of a rose, which is
accompanied by the delight to our nasal
senses. But the rose floats on its own tiny
astral pond, which is invisible to our low
tuned consciousness. The pond reflects the
beauty of the rose at a higher frequency to
our physicality. The pond 'is' its higher
beauty. The surface of the pond reflects, as
would the surface of a material existent
pond. But the astral pond surface is all
directional. Imagine the surface of a
tranquil woodland pond. It glitters with the
reflected sunlight. Imagine that pond
surface tilted and extended through every
level of the physically perceived space. The
pond surface would quickly form into a
sphere. A 'solid' sphere, for every aspect
of the space within it would be occupied by
a surface of the reflective pond.

The pond would then reflect every glittering aspect of the sun that was shining upon it. Every material form has its astral pond. Nature has the wisdom to tune down our senses to such experience overload, letting them evolve at a qualified rate. The enlit, being the servants of Nature, adjust the evolving according to the ultimate will.

Active ghosts, sentient life shadows, light bulbs that won't acknowledge they have been switched off, quickly learn how to use an objects astral pool surface. They are at a level to perceive that surface, and by perceiving they can activate the will to control that surface. They can motivate the active will to 'take hold' of the surface. They can grip the astral mirror sheet of reflection. If the will is aggressive or frustrated this results in what is perceived at material conscious level as poltergeist activity.

Active sentient life shadows can take hold of an objects astral pond surface and move the object through our perceived space. We perceive it as something supernatural. A fish within the pond perceives your hand entering its pond space to rearrange the pebbles. Is your hand supernatural? The fish will think so. But by having such thought gives reality to their idea of an event which is supernatural. Supernatural is only an idea that needs to be clarified.

Active sentient life shadows remain sentient beyond physical existence due to the weight of thought they have placed within gravity during that physical existence. All material life leaves such a gravity print. When the material carrier expires the gravity print remains with an echo of the life transient consciousness and will. Even a flower will leave a 'ghost'. But it will only exist for a fraction of time. The flower's 'thoughts' are held within its tiny gravity print.

These will almost instantly dissipate into the greater thought. The process applies throughout the life spectrum up to human consciousness and will. Our gravity print lasts much longer as a sentient containment field, vibrating close to perceived physical level. But the consciousness dissipates higher after a few days, before it has had to chance to reason on its new state. The essence then moves on beyond the pull of perceived physical matter and the will to control becomes rarefied. Here the life essence will stay for some considerable time, as we experience it, but they can still be reached by those with the gift of trans-materiality.

However, in some instances the thought generated while in physical life can be strong enough to leave a form of gravity print that remains fixed for much longer then the universal parameters. A glowing light bulb is smashed, but the light within it is so bright that it refuses to vanish.

Even when the glass is broken, and the
filament snapped, the light remains fixed in
the shape of the bulb, continuing to light
up the room. Eventually it accepts that it
is no longer a physical light bulb. The
light will then vanish into the static of
the radio waves that fill the air.

It is the same with sentient life shadows.
Sudden or violent death can often wrench the
physical carrier away from its gravity
platform to leave the conscious will
suspended within its resonance template.
Gravity holds everything together across
space and time. Part of you is holding a
star together. Gravity is also the pond on
which thought floats and in which conscious
will swims. Gravity is the Universe for the
soul and the spirit. If all these elements
are strong enough, gravity will retain the
memory of them for much longer and the
sentient entity remains beyond its physical
carrier. Material astral resonance control
is quickly learned.

This results in classic 'ghost' activity.
For those of tragic passing it can be an
expression of anger, or the need for
revenge. For those of natural passing it can
be expressed as a reassurance to loved ones,
or as a need to have some mischievous fun.
You see that beautiful sports car. You jump
in. But do you want to sit there, knowing
your powers? No, you want to drive the
thing. You can move an object by rippling
its astral resonance and then scare us. You
can turn the ignition key, so you turn it.
You move the vase to fall off the shelf.
Some do have will control. It is a form of
blessing. Their shine was already closer to
the greater shine. But most take the joy
ride. Ghosts are only ex human.

Again, Nature provides a simple analogy.
Look at a light bulb for a few seconds and
then close your eyes. The image of the light
bulb remains, 'remembered' on the receptive
cells of your eye. You are still consciously
aware of the light.

Eventually the cell light memory will fade to become part of your actual thought mind memory. Then you will remember seeing an image of the light, that was itself an after image. Imagine that the first after image actually had a sentient consciousness. It was a 'living' light bulb after image 'remembered' on the receptive cells of your living eye. This is an analogy of a 'ghost'. It is also an analogy of our relationship to gravity and the cosmic mind. In this case 'we' are the 'living' light bulb after image 'remembered' by the cells of a much greater eye.

As for other forms of ghost they are as gravity c.d. recordings, played back through the laser recording head of your mind when play back conditions are suitable, or sometimes when the weather is suitable. Question: Is a reportedly haunted house haunted when it is empty? Answer: No. Such recordings need a physically enclosed mind to manifest through.

Can a c.d. play without a c.d. player – NO.
If a c.d. had the super will to place itself
within a c.d. player, could it switch the
player on in order to play itself? NO. It
would need a higher consciousness YOU to do
that magic. And as for the jolly sentient
life shadows, would they bother haunting if
nobody was there? You could approach an
assured empty haunted house and then hear
some activity within. But you are within
proximity to the house and they know, for
you have willed to approach the house. You
are also within range for the recordings to
activate. Put the house on the moon, then
you would be truly out of range, until you
willed to go to the moon.

The supernatural isn't. There is nothing
supernatural within the Universe, for the
universe is the most natural thing that
exists. And we exist within it.

If something was supernatural it wouldn't be
natural for us to experience it.

If it isn't natural for us to experience,
then there is no point in its having an
existence. The Universe naturally does
everything to have existence. It has no time
for the supernatural.

Octaveiates of the Enlit

Life is the distillation of Universal will
into self awareness.

Everything is Life.

Life exits to be all things unto itself.

The will of Life is to perfect its dream
through Universal experience.

Light is the shadow of all gravity.

Gravity is the shadow of all thought.

Space is a factor of thought.

Imagination is the music of thought.

Music is an illumination of imagination.

Art is imagination's will for self
awareness.

All language is a resonance of thought.

One can learn the language of a star through
a single heartbeat.

Magnafactal Octaveiates

True magic is the art of controlling the
resonance language of all temporal matter.

A drop of water contains the Universe.

All answers lie within one question.

The question to all answers can be seen all
around the one seeking the question.

One exits to ask the question.

Consciousness is self reflecting thought.

Mind is the gravity of existence.

It is my hope, and, prismised through me,
the hope of the enlit, that you now shine
brighter through seeking these fallen jewels
of wisdom. It is hoped that they have served
the purpose for which they have descended.

The illusion of light transmits the reality
of a true greater light. To dream is to
exist. The light that we perceive is the
reflection of our dream. We dream to dream,
and by dreaming we create. We create the
existence in which we dream.

WordLight Twenty.

Not forgetting all sweet words said,

Steeper grows the way ahead,

Yet still we climb to reach our star,

So longing for dream's reservoir.

As the light we quest seems out of reach,

Words of wisdom strain to teach,

Sparking the inner voice to speak,

Loosing no wit nor will to seek.

If then at last, the answering gains,

Stand to look at what remains,

And if it pleases, leave it so,

If not, then ask again, and future know.

That light trickles through our infant eyes,

Through tears of wonder, failing to

disguise,

The fear that we might look too deep,

Finding secrets that we cannot keep.

And if all of this shall be our test,

To fail, or climb above the rest,

Of Nature's kingdoms, spread about,

Through it all,

Our light must shine, and never once go out.

Light tells us we are dreaming, and through

us, tells itself of its own dream.

The enlit reach and shine and in so doing light up their own reality. For we are their dream as they are ours. By reaching they spread their own self awareness. They create new thoughts through us. And new thoughts create new realities. They are as real as reality allows them to be. Their reaching strengthens that reality and, in turn, their own reality.

The enlit exist, and the proof is to have the thought that they exist. It is their thought. Our minds are the mirrors for their own questions. They ask of all pasts and of all futures. Through us they can see the way they have travelled. They steer from all darkness that had once befallen. The thought of them through us is the star to follow.

The thought of them will take us to the reality of that thought.

And on reaching that time of the reality of
that thought will be the time of knowing the
reality of what we shall then be. It is then
that we shall reach for others to join us.
The enlit exist for we exist. We are our
past and future one. Our thought creates us.
Our dreams are shared. Our time is reached.

Not the End.

As Nothing Ends.

As Nothing Ends Nothing Begins.

It Always Was.

R.O.A.M.

LiT Gleanings.

A Shine on the Universe.

The expansion of the Universe is an illusion partly due to our size in relation to the cosmos. We are so small. The motion that we see appears as an expansion, but we are only seeing a tiny fraction of the true motion. Nature provides the answers to the big questions and gives them in very small ways all around us. We simply have to look down occasionally.

Imagine a large clear pond of smooth still water. You throw a small pebble into the centre. There will be a splash and then ripples will spread out across the surface of the smooth pond. To an imaginary pond surface atom being the substance of the pond, the atom being's Universe, would appear to be expanding. And it would appear to be expanding from a central point. The atom being could deduce that his Universe began with a big bang. But, from the greater perspective of someone on the grassy banks of the pond, the expansion of the atom being's Universe is seen only as a ripple on the surface of the pond.

The ripple itself is a travelling distortion of a section of the pond substance, its surface. The pond itself is not expanding. When the ripple has passed, that section of pond is still there. It hasn't moved. We are so small compared to the Universe that we only see an atom's slice of the Universe pond surface.

We see only one aspect of motion that we perceive as expansion because we cannot see the whole pond with grassy banks. We cannot realise that the pond has another depth to it. Our viewpoint is so low as to see only straight across the surface of the pond.

When the ripples hit the grassy banks of the pond they dissipate and reflect back weakly, interacting with the oncoming ripples. Eventually the pond surface will become smooth again, until someone decides to throw another pebble. Regarding the Universe, the perceived expanding motion of star matter continues until it reaches the Universe grassy banks of the gravity influenced curvature. The 'ripples' will then continue 'out of our sight' over our perception horizon line, until they return 'behind' us. But there is no behind us. Behind and in front are reflections of a single central perception of mind-space. Both directions are the same. All directions are reflections of consciousness. Movement is a thought.

It is this thought sameness that allows the 'returning' Universe ripple to appear as if it actually expanding. In all directions we see an illusion of expansion but in reality it is returning to us. It is returning to an 'us' that we cannot perceive because it is an 'us' that lies over our mind perception horizon. The pond surface atom being cannot step out onto the grassy banks. He cannot see that the surface of his pond is only a thin slice of its true depth.

The atom being will also devote a great deal of thought to the perceived central location of the expansion. There will be the dramatic appearance of some great event that caused it all. It will be a big bang. However, if he could step onto the grassy bank he would see that his perceived expansion is only a ripple on a great multi directional mass. He would see that the big bang event is only a temporary anomaly caused by someone throwing a pebble. You.

The central point of his big bang theory is constantly at a new location. The pond is situated on the surface of a planet that he had no idea about. The planet is rotating. The planet is rotating in an orbit within a solar system that he had no idea about. Already he has overload. The solar system is situated at the outer edge of a galaxy that he had no idea about. And the galaxy is rotating. He would see that his expanding pond surface Universe is expanding from a multidirectional point of motion. The expansion isn't expansion because it is moving in all directions within the same time frame.

The stars and galaxies that we perceive with physical instruments are 'moving away from us' in all directions. The great something that it is moving into is in multi-motion. Our minds are geared to seeing just one plane of this motion. This is Nature and the Universe designing reality so that we do not get perception overload.

This is why our eyes only register a small bandwidth. But our minds have learned about X-rays and microwaves etc. This is our awakening. Our greater Universe sight has to evolve slowly and the mind radio frequency has to be turned up very gently.

The pond surface atom being will now see a new reality to his big bang theory. The central location of the pond, where it was all supposed to happen is seen as – a pond. He will wonder where this big bang came from, if there was one after all. But how did the expansion illusion start? The atom being's mind might not stand the shock of you approaching and explaining that you threw a pebble and started it all. And then another shock of being told that you are thinking of throwing another one. Your pebble throwing thinking might explode his thinking. It was you who had the thought to throw the pebble and create the ripples, confusing the tiny atom beings. There was no big bang, but there was your thought.

The Universe did not begin with a big bang. Nothing just begins. There has to be a reason for anything to begin, and that reason is a smooth constant. There is no thought reason for a messy big bang. It is against the essence law of Nature for something to just begin. Before that beginning would have to be something. Something always was and always will be. Matter is in a constant state of space-time transition, from one level of form to another. The Universe is happening because it wants to happen and has always wanted to happen. It lets us see a part of it that it wants us to see, because we are not yet ready to face the mirror fully.

The Universe knows what it looks like, just as you do, for you have seen yourself in a mirror many times. But you can play a game with yourself, as does the Universe.

You can turn your back to the full length
mirror. You cannot see your image, but your
mind knows what that image looks like. It
has imagination. Now you raise your hand and
turn the palm to face the mirror behind you.
You are letting your hand see itself. There
are no eyes in you hand, but your mind is
within your hand, to turn it. It did not
turn itself. Your mind is focused on your
hand and through your mind you can see
'imagine' your hand in the mirror. The
Universe toys with us by allowing us just
tiny glimpses of the whole mirror.

To the cells that make up your hand you are
their Universe. If they had a slightly
higher awareness they would look in the
mirror and see themselves as a hand. But
they would not know what the great shape was
that was turned away. They would not know
that the great shape's greater mind was
choosing to turn away and only allow their
hand cell minds to look at themselves.

As they continued to look they would become aware that they were connected to this great and mysterious shape. They would eventually realise that it was their Universe. They were seeing their Universe and they would be happy. But they wouldn't know that it was turned away. They wouldn't know that it was turned away because it wanted to be turned away. Suddenly, the hand would see something new. It would be given a higher awareness. It would see that there was another hand. Eventually you would turn and allow yourself to see yourself fully, again, as you have done many times. And then you would play the game again. You would throw another pebble into the pond. If you wished to see a reflection even closer to thought reality you could decide to place a second mirror in front of the first mirror. You would then see both sides to yourself at the same frame of time. The Universe of you would be mirroring the greater Universe of which you are a part.

It can choose to see every side of its creation. But this is too simple. It is too easy. It is too perfect.

The Universe likes to work. It likes to create, both positive and negative. Even destruction is negative creation. The Universe's Nature laws always retain the balance in favour of positive creation. But too much positive creation would lead to the possibility of perfection. That possibility exists, but it cannot be allowed to become a reality. Perfection would be an end state.

The Universe cannot allow itself to exist at an end state for then it would have no reason to exist. It would be an end to itself. This means that nothing would follow, and nothing is the only true impossibility. Nothing exists as a concept of nothing within the mind. The Universe is a prism for all levels of mind so it cannot fall into the true impossibility because mind will prevent it.

That is why nothing is the only true impossibility, because it exits as such a concept. Nothing is impossible. These words are proof. The Universe must constantly correct the drift towards perfection by controlled negative creation. You are a runner, training for the gold medal speed. The Universe of your body will at some point strain a muscle. Nature will apply its law.

The Universe began as it now continues. It continues always. It is always beginning for it is always recreating itself. The Universe did not begin with a singular event. A singular event would itself have to have something for it to be a singular event in. That something itself had to have something for its origins to event from. The Universe did not begin with a singular event because it is already a continuous singular event, and always has been.

Nature has engineered our human brains and mind to accept time and space and material motion at a set perceived rate. It is the rate that we all share to within fine parameters but it can be artificially modified with stimulants. Nature has set the perception rate at different levels for other forms of life, according to their awareness need.

A plant's Universe is totally different to ours. It does not need to know how to drive a car or use a computer, but it will be far more in tune with the planet's energy flows, for its nourishment is closer to source. Most of our nourishment is processed for it to acceptable. We use the sun for energy and recreation, and study its properties through physical instrumentation. Plants 'know' the sun. They have to. It is within their design concept. They know the sun, the moon, the planets, and the stars at a level that is beyond our perception, for their existence needs to have such knowing.

The plant is at a much more fundamental
level with the energy flows, especially from
the sun. Plants will know when the sun is
having an energy surge, before our physical
instruments tell us. The sun will tell them,
in a form that they can understand, and then
it tells us, in a form that we can accept.
We gain energy from the sun but our physical
existence does not require plant and animal
perceptions. Our minds are wider, yet Nature
has to close some doors to allow it. But it
has the wisdom to retain the option to open
them. A plant has the skill to talk to the
sun, and has awareness of us. But the plant
could never build a space ship, while it
remains a plant. Yet we retain the doors to
have plant and animal awareness. They are
locked but not permanently. They can be
opened with specialized training and
techniques, for such extra light requires
great care. Our minds are set in destiny to
face the Universe. We have reached that
point. It is new to us and bright.

Nature, in its wisdom, has provided us with sunglasses. The time to remove them is still beyond the horizon, but it has gained pace. We are becoming aware of the quickening.

A plant could never build a space ship, while it remains a plant that we recognise as a plant. But if the plant was genetically modified and its mind widened, it could be raised to our level. Finger tendrils could be designed. Eyes could be created, which would mean that the plant would need a true brain. The concept could also apply to animals. However, they would then not be true plants or animals. A cat with paws that could play a piano would not be a true cat. The Universe of our human form and intellect would have imposed itself on their Universe to give them a new one. Such possibilities have now entered the dream boundary of our technical capability.

Modifying plants and animals would also modify their time perception. The cat would need a totally different concept of time in order to play the piano. True cats need the time and space awareness to judge when to jump to catch a mouse, not to get the pace right for Debussy's Clair De Lune. Their time perception would change to meet their needs. It would be lifted to a level close to ours, their creator. But our level of time and space perception is itself geared by Nature to fit our own current needs.

Time and space behave according to how they are perceived. They are actors on a stage and play a part to suit the audience. Time does not move. This is an illusion created by Nature to keep us happy. Matter moves and is in a constant state of multi-motion and transformation. This multi- motion and multi-transformation needs a stage.

It needs something to happen in. Time and
space are automatically created to be its
stage. Without motion there would be no time
and space. There would be no need for it.
The Universe would be a nothing solid. But
nothing is impossible. The tiniest amount of
motion would create the time and space for
that motion to happen. It would build the
theatre. No particle of substance could ever
stay still. All particles are in constant
reaction to other particles. Without such
reactions the functions of matter could not
continue.

Such reactions create time and space. Our
minds are tuned to perceive this as a steady
flow. We dream it as a stream. We have
memory travel of the past and imaginings of
the future. This travel is an event change
and matter transference travel. Time is a
phenomenon of consequence. We say it takes
time to do something. We need more time. We
say it takes a certain amount of time to
travel from one physical point to another.

It takes energy and space to do something.
It is the needing of more energy and space.
It takes energy and a certain amount of
space to travel from one physical point to
another. The time element is a consequence
of the rate at which such energy is used and
the amount required. Give something more
energy and you reduce the time for the
event. When our quantum brains receive more
positive energy, time is enhanced. With
negative energy it can be enhanced and
distorted. The doors can loosen prematurely.

Time is a consequence. It exists as a state
of Universal mind that is giving itself the
space to happen. You move your hands to the
computer keyboard. You move them through
space with the will to use the keyboard and
create the time to move them towards the
keyboard. Without the action of your will
that fraction of time would not have
existed. It would not need to exist for
there was no will for an event.

Every particle creates the time for its
motion and collectively they create the time
that our mind perceives as a flow.

To travel through time would not be as
travelling back and forward, it would be as
a decent and an accent. It would be as
movement on a reflection between the two
mirrors. It would be a movement of state,
not a physical movement. You would 'move'
from one state of time to another, from one
frequency of time to another. Nature always
provides an analogy to every question. We
ask questions about the stars and the
answers can be found at our fingertips. Time
is as the information on a film disc. The
images on the screen are the moments of
reality that we experience as a single
continuous now. But the information for all
the screen images are on the disc – in the
image past – and in the image future.

The laser beam that reads the information

gives us a tiny instantaneous fraction of

the whole film that we perceive on the

screen as a moving image. The laser beam of

our mind reads the information of space-time

in a similar way but on an infinitely higher

level. The disc information and laser reader

are an ox cart. Our minds are a star ship.

We can skip scenes back and forth. But there

is no actual movement. We are selecting a

different pattern of disc information. Time

travel is turning back the information,

tuning down, or tuning up to information

frequency that has yet to appear on our mind

screens. The Universe has its film already

stored on its own disc of matter memory and

matter anticipation. The Universe knows what

it was and what it will become. This knowing

is its disc. It is its dream. The secret of

time travel lies in asking the Universe to

allow us to enter its dream. We have been

given the gift to have a mind to discover

the way of asking, for then the Universe can

discover itself, as it must do continuously.

The Universe must discover and know itself through all life forms. All life forms are itself.

The secret to all things lies in asking the Universe to enter its dream. It is the path to true magic. The dream runs through all matter, illuminating it with a tiny essence of Universe mind. All matter then shines with mind. It becomes the truth of the Universe. All matter is life.

T. HuGotH.

R.O.A.M.

R.O.A.M.

Notes.

Alex and I have known each other for thousands of years. We have only recently discovered this.

When the ramblings finally came together in a coherent form I needed someone who could act as a catalyst and whom would understand my wish to stay shaded. The forces of fate brought us together. Knowing my interest in such things friends began to mention Alex, for others had mentioned him to them. I began to have an enlit awareness that this sudden news of a hypnotist had a purpose, especially when I discovered that we lived in the same area. This awareness became positive when he started visiting the pub I use regularly.

For him it was just a decision to try the
pub for a change. He didn't know he was
being guided. Our circle of friends mixed
but didn't tie, and I stayed on the edge.
This happened for a number of occasions
until the spark lit. Alex was sitting close
and overheard my conversation with a friend
about my inspirations. He enquired and he
was interested. That was it. He realised why
he had been drawn to the pub and I realised
he was more than the catalyst I had asked
for. He was a prism. He realised the
inspiration needed to shine wider. We soon
discovered just how well we knew each other,
and for how long. It was a destined meeting.

*****

I have never seen a ghost, but I have met
people who have, and I have family whom have
experienced such events. My only experience
was as a youth, quite a few years ago. I
slept in the back room of our terraced
house.

Late at night, at the same time, I would hear distant scratching from the yards at the back. At first I thought it was a neighbour's dog. And then I realised that the neighbours in the direction of the scratching didn't have a dog.

The scratching was heard always at the same time, around one o'clock, and it would only last for around fifteen – twenty minutes. It would wake me up. Then I got up and looked through the window. The scratching definitely came from a direction where there was no dog. I went down into the yard. The scratching could not be heard. At that point the scratching took a turn. It began to get closer. Soon, the scratching could be heard from the yard next door. They had three cats. I'm not sure now why I allowed it to continue for so long. Perhaps I couldn't accept what my subconscious was telling me. Eventually the realisation broke through, when the scratching started to be heard from our own yard, and then from the wall brickwork. I bought a small plain solid silver crucifix, with a silver chain.

I had it blessed and I placed it on the T bar of the window frame. From that moment the scratching stopped. The silver crucifix stayed there and I only heard the scratching once again, several weeks later, as if from way down the yards outside. I never heard it again.

Since then I have moved home a number of times. They are events that are now old history, but I still have the crucifix.

*****

Ghosts etc are a facet of Nature that has yet to be clarified. They are Holographic Interactive Trans-dimension Sub Entity Manifestation. H.I.T.S.E.M. Our deep materialism is a hindrance to our understanding of hitsems. It puts weight on our mind. It reduces our greater Nature vision. Materialism is a true necessity that needs to balance with such vision.

Ghosts and related manifestations are a form of reality that is as real as the shadow that we cast on the street.

You cast a shadow from the light of the sun. In a similar way you cast a shadow on time and space. The shadow on time and space includes an essence of you. That shadow can hover, and often does. Imagine that you walk down the street, but your shadow remains on the time and space of the pavement.

You had the life force to create the shadow. The life force moves on but the shadow doesn't want to move just yet. It was happy on the pavement, or it hasn't yet drawn on the pavement the crayon sketch that it wanted to. It becomes a hitsem.

*****

So called U.F.O.s are also an integral part of life, exactly as are hitsems. There are many forms. Some are a form of space ghost. Some are life entities in their own right. Space fish. Some are actual vessels, occupied or robotic. These are close to our level of technology. Thought probes. Time and dimension ships. Earth lights. Higher spirit entities.

The enlit have advanced to being fully
trans-galactic. Their physical forms still
exist, on suitable planets within many
galaxy outposts. Their appearance would not
shock us unduly. We have already been tuned
to accept such an image with reasonable
ease. My imagination is allowed to see them
as tall, matchstick and humanoid. They are
graceful and they glow softly. Their eyes
are star pools. They concentrate on galaxies
where life is developing technology. They
communicate in many ways. Their minds can be
sent into any living form. If a flower
inspires you, that could be a communication.
They seed thoughts and ideas which can lead
to music, art and science. They do not write
the words of a poem, but they can create the
thought for them. And they can create the
thought of a name that is appropriate. The
enlit. In this way they gently nurture. It
can take centuries but this is not a problem
to them. They have mastered space and time.
Their technology has advanced to a level we
would call magic.

It is on their horizon when their physical form will become redundant. Still several million years, but to the Universe this is within the next few seconds. They have reached the level where they can travel through time to see their next existence level. They are already prepared. They will become star beings and help the next enlit, who will be helping the next advancing life.

*****

The pyramids are calling cards, saying, we have reached a level were we can do this. It is a structure chosen because it can only be built using advanced techniques of thought gravity control. The skill had been handed down from previous advanced civilisations, who in turn had learned or improved their own skill from contact with other planetary civilizations. Aliens did not build the pyramids, but they were built with skill descended from them. The blocks of stone were in a form of suspension and then physically manoeuvred into position.

They did not have the full skill to mentally tell the blocks where to position themselves. The skill had been diluted over the centuries. The stones were influenced into suspension and then the structure was built to a pattern. They did not use a long ramp, or a ramp that climbed around the pyramid as it was being built. How would those ramps stay in place on a sloping surface with massive stones being dragged over them? The stones were not dragged. They were levitated. Ramps as such were not needed. But they were built to a strict pattern. The levitation was only a few inches at the most and the blocks had to be guided with great precision, still using rollers and levers and a form of crystal ball bearing.

They were built to leave the power of the inspiration for future generations. They are a thought store to keep a spark of wisdom alive.

*****